D0046286

ALSO BY STEPHEN M. KOSSLYN

Image and Mind

Ghosts in the Mind's Machine

Wet Mind: The New Cognitive Neuroscience
(with O. Koenig)

Elements of Graph Design

Image and Brain

Psychology: The Brain, the Person, the World
(with R. S. Rosenberg)

Fundamentals of Psychology
(with R. S. Rosenberg)

The Case for Mental Imagery
(with W. L. Thompson and G. Ganis)

Graph Design for the Eye and Mind

Cognitive Psychology (with E. E. Smith)

Clear and to the Point

Abnormal Psychology
(with R. S. Rosenberg)

Better PowerPoint

ALSO BY G. WAYNE MILLER

The Work of Human Hands

Coming of Age

Toy Wars

King of Hearts

Men and Speed

The Xeno Chronicles

An Uncommon Man: The Life and Times of Senator Claiborne Pell

Thunder Rise

Asylum

Summer Place

Since the Sky Blew Off

Vapors

The Beach That Summer

TOP BRAIN, BOTTOM BRAIN

Surprising Insights Into How You Think

Stephen M. Kosslyn, PhD,
and
G. Wayne Miller

SIMON & SCHUSTER
NEW YORK LONDON TORONTO SYDNEY NEW DELHI

Simon & Schuster
1230 Avenue of the Americas
New York, NY 10020

First Simon & Schuster hardcover edition November 2013

Designed by Ruth Lee-Mui

Manufactured in the United States of America

1 3 5 7 9 10 8 6 4 2

Library of Congress Cataloging-in-Publication Data

Kosslyn, Stephen Michael, 1948-
Top brain, bottom brain : surprising insights into how you think /
Stephen M. Kosslyn, PhD, and G. Wayne Miller.
pages cm.
Includes bibliographical references and index.
1. Cognition. 2. Brain. 3. Cognitive neuroscience. 4. Cognitive psychology.
I. Miller, G. Wayne. II. Title.
BF311.K674 2013
153.4—dc23 2013024491

ISBN 978-1-4516-4510-1
ISBN 978-1-4516-4512-5 (ebook)

To Robin, Justin, David, and Neil—
all of whom have taught me a lot about modes of thinking.

—SMK

To Mary and Duke Wright,
my biggest supporters from the very beginning.

—GWM

Contents

Preface

Why Another Brain Book?

This could be the golden age of books about the brain.

Visit a bookstore or go online and you will find a dizzying selection of volumes exploring the role of the brain in work, relationships, creativity, emotions, personal fulfillment, and more. Neuroscientists, psychologists, life coaches, and others weigh in. Theories and insights are offered. Authors provide guidance about how to enhance your knowledge, skills, and abilities and how otherwise to improve your brain performance—with the promise of a better life, in one way or another.

So the question becomes: Why one more book about the brain?

Because this one is different.

Top Brain, Bottom Brain offers a new, defensible theory of brain functioning and psychology, based on a major anatomical distinction that is frequently overlooked. The "Theory of Cognitive Modes" that we present here is built on conclusions arising from decades of research that, for the most part, has remained inside scientific circles. A "cognitive mode" is a general way of thinking that underlies how a person approaches the world and interacts with other people. To our knowledge, this is the first time that these findings have been systematically brought to a mainstream audience.

We have learned a lot from mistakes of the past, in particular those whose origins can be traced to another way to organize the brain: into its left and right halves. *Top Brain, Bottom Brain* debunks this dominant mainstream brain story of the last half century, the story of the alleged great divide between the "analytical/logical" left and "artistic/intuitive" right halves of the human brain. Although the left and right halves of the brain do have some different functions, they are not those described in the commonly accepted story. In the service of not repeating the mistakes of the past, we briefly consider how this story came to be so broadly embraced by the popular culture—and why it is scientifically unsound. Our theory does not fall prey to the problems that beset this earlier one, in particular because we do not try to develop a simple black-and-white dichotomy to describe how large portions of the brain work. Rather, we emphasize the role of different systems that are implemented in different brain regions.

We intend this book to be accessible to the general reader, but we also hope that it will influence the scientific community. Since neuroimaging (also called brain scanning) burst onto the scene in the mid-1980s, many studies of the mind and brain have become noticeably less theory-oriented. Although neuroimaging is a valuable tool in modern neuroscience, it is not always used very effectively. Some researchers now simply ask people to perform a cognitive task (such as playing chess or thinking about a particular topic) while their brains are being scanned—and then see which parts of the brain become activated. This approach is radically different from the traditional one, in which the researcher tests specific theories. The traditional approach is important because science makes progress by devising increasingly better theories, and thus it is essential that theories be developed and evaluated. In this book, we seek to return to the hypothesis-driven tradition by proposing a new, plausible, and coherent theory that is strongly rooted in empirical findings.

We remind the reader at critical points that we are working with a theory, and that rigorous empirical tests of many key predictions have yet to be conducted. We hope that the reader will take the ideas we present as well-founded hypotheses that are worth considering, not as received truth. We also hope that this book will inspire a new round of studies that will further enrich our understanding of how the brain really works.

More than that, we hope that our Theory of Cognitive Modes will be an important contribution to understanding ourselves, as humans and as individual people. The theory we offer here should prompt you to think about issues you probably did not previously consider. Whether your interest is to evolve personally, socially, or in business—or all three—we believe that understanding and considering the Theory of Cognitive Modes can benefit you. We hope you find this book thought-provoking. May it lead you to useful insights about yourself and the people in your life.

To quote the ancient Chinese philosopher Lao Tzu: "He who knows others is learned / He who knows himself is wise."

Stephen M. Kosslyn, PhD
San Francisco, California

G. Wayne Miller
Providence, Rhode Island

TOP BRAIN, BOTTOM BRAIN

Chapter 1

A New Way of Looking At What Your Brain Says About You

The human brain is arguably the most complex object in the known universe. By one estimate, the number of possible combinations of connections among its many billions of cells rivals the number of elementary particles in the universe.[1]

So how do we begin to understand such a complex organ?

This is no mere academic exercise. Our brains underlie everything we do—from waking up in the morning, to navigating the physical, social, and emotional shoals of the day, to falling asleep, perchance to dream. Collectively, our brains organize enterprises and societies.

But the fact that the brain is extraordinarily complicated does not imply that it is forever incomprehensible. The field of psychology would not exist if that were true. In this book, we present a new way to look at the brain, which may help you to understand how your own brain gives rise to your thinking, feelings, and behaviors, and how it affects your relationships with others. Much of the science on which the theory is based has gone largely unnoticed outside research centers.

In spite of the brain's complexity and the relative youth of the field of neuroscience, two insights allow us to begin to understand how the organ functions.

First, it is possible to examine any object—including a brain—at "different levels of analysis." Take the example of a building. We can consider it in terms of its architecture: its floor plan, the shape of the rooms, where the doors and windows are located, and so on. We can also go a level of analysis deeper and consider the materials used to construct it: brick, wood, plaster, and so forth. And, if we are so inclined, we can go down yet another level and consider the molecular structure of the materials, such as the arrangements of atoms in a brick.

Which level is best? That depends on the question we want to answer.

If we want to know whether the house will have enough space for a family of five, we want to focus on the architectural level; if we want to know how easily it could catch fire, we want to focus on the materials level; and if we want to engineer a product for a brick manufacturer, we focus on molecular structure.

Similarly, if we want to know how the brain gives rise to thoughts, feelings, and behaviors, we want to focus on the bigger picture of how its structure allows it to store and process information—the architecture, as it were. To understand the brain at this level, we don't have to know everything about the individual connections among brain cells (neurons), about the way ions pass through cell membranes, or about any other biochemical process. Nor do we need to know the structure of the molecules that compose the interior parts of neurons. Researchers in neuroscience and psychology have learned a lot about how the brain works at an architectural level of analysis even though they do not fully understand it at lower levels of analysis. This book is built on much research conducted at this high level, from a macro more than a micro view.

The second key insight that allows us to begin to understand how the brain functions is the fact that—at the level of architecture—the brain is organized into systems. And these systems themselves are organized into subsystems.

For example, one postage stamp–sized area of the brain, known as V5, plays a crucial role in processing the visual perception of motion. (V stands for visual, and 5 indicates the area's remove from the first visual area, V1.) When a person sees a moving object, neurons in this area fire; if Area V5 is damaged, the person has difficulty visually perceiving movement. Area V5 is part of a larger system that registers the locations of objects. And this larger system is in turn part of a still larger system that coordinates the location and identity of objects.

The point is that, even if we stick with one level of analysis (in our case, how information is stored and processed), we can study small, highly specialized areas or we can study larger areas. Moreover—and this will prove crucial—as we consider increasingly large tracts of brain real estate, we need to characterize what they do in increasingly inclusive terms: terms that encompass all the more specialized things the constituent areas do.

In this book we rely on both of these insights: We use a relatively high level of analysis, akin to architecture in buildings, to characterize relatively large parts of the brain.

Our examination would not have been possible even a few decades ago. Those of us interested in the brain are fortunate to be living in the present era, when neuroimaging techniques have been invented and refined. Such techniques are well suited for examining how large parts of the brain store and process information, and much has been learned from the use of such techniques—in combination with older ones, such as examining how thinking, feeling, and behaving are altered following damage to specific parts of the brain.

To explain the Theory of Cognitive Modes, which specifies general ways of thinking that underlie how a person approaches the world and interacts with other people, we need to provide you with a lot of information. We want you to understand where this theory came from—that we didn't just pull it out of a hat or make it up out of whole cloth. But there's no need to lose the forest for the trees: there are only three key points that you will really need to keep in mind.

First, the top parts and the bottom parts of the brain have different functions. The top brain formulates and executes plans (which often involve deciding where to move objects or how to move the body in space), whereas the bottom brain classifies and interprets incoming information about the world. The two halves always work together; most important, the top brain uses information from the bottom brain to formulate its plans (and to reformulate them, as they unfold over time).

Second, according to the theory, people vary in the degree that they tend to rely on each of the two brain systems for functions that are optional (i.e., not dictated by the immediate situation): Some people tend to rely heavily on both brain systems, some rely heavily on the bottom brain system but not the top, some rely heavily on the top but not the bottom, and some don't rely heavily on either system.

Third, these four scenarios define four basic *cognitive modes*—general ways of thinking that underlie how a person approaches the world and interacts with other people. According to the Theory of Cognitive Modes, each of us has a particular dominant cognitive mode, which affects how we respond to situations we encounter and how we relate to others. Of course the details matter a lot, but these are the basic ideas we develop here.

Systems, Not Dichotomies

In this book, we use what researchers have learned to present a new theory of brain function that hinges on how the top versus bottom parts of the brain interact. But we do not try to characterize the top versus bottom parts of the brain in terms of a simple dichotomy or set of dichotomies, which was exactly what was done with the existing and well-known division of the brain into two halves: namely the left versus the right, the dominant pop-culture brain story of the last few decades. You have probably heard of this theory, in which the left and right halves of the brain are characterized, respectively, as logical versus intuitive, verbal versus perceptual, analytic versus synthetic, and so forth. The trouble is that none of these sweeping generalizations has stood up to careful scientific scrutiny. The differences between the left and right sides of the brain are nuanced, and simple, sweeping dichotomies do not in fact explain how the two sides function. (We will return to this topic in chapter 5.)

Given the need to characterize large parts of the brain with inclusive terms, it is tempting to invoke simple dichotomies. The problem is that such dichotomies can't even begin to capture how large parts of the brain really work: When we look at big swaths of the brain, we necessarily group together lots of smaller areas that do specialized tasks, such as motion detection. And as the region we examine is larger and larger, the smaller areas that are being grouped together will be more and more diverse. Even if some of these very specialized areas can be described by a dichotomy, no single characteristic may unite all the small, specialized areas—the small areas may not have a single function or functions in common. A large region of the brain may be likened more to a rope than a wire: In a rope, small strands overlap, but no single strand runs through the entire thing.

When considering large portions of the brain, we need to think about systems—not dichotomies. A system has inputs and outputs, and a set of constituent components that work together to produce appropriate outputs for particular inputs.

A bicycle is a familiar system: The inputs are forces that push down on the pedals, slight movements of the rider's body made in the act of balancing, and force that moves the handlebars. The components include the seat, the wheels, the handlebars, the pedals, the gears, the chain, and so forth. The outputs are the bike's forward motion, keeping upright, and going in a specific direction, all at the same time. Crucially, the components are designed to work together to produce appropriate outputs for the system as a whole—for the entire bike. The handlebar is connected to the front wheel for steering, the seat is over the pedals to make it easy to push down, the gear chain connects to the rear wheel to cause it to propel the bike forward, and so on.

The same is true of the brain: It has different areas that do different things, and the result of the brain areas' working together is to produce appropriate outputs (such as your avoiding an object) for particular inputs (such as specific sights and sounds). For instance, if you see a car roaring toward you, you jump out of the way.

Top Brain, Bottom Brain

The Theory of Cognitive Modes that we develop in this book is based on organizing the brain into two major parts, top and bottom—each of which we will characterize as a large system that processes information in particular ways. As we show, we gain a lot by organizing the brain into these two large systems, noting how constituent parts work together. Let's begin by being clear about what we mean by the top and bottom parts: Look at the diagram of a side view of the brain, which shows the cerebral cortex, the thin outer covering of the brain where most of the bodies of neurons reside. The cerebral cortex is where most cognitive activities arise—and in this book we focus almost entirely on the cerebral cortex (not the "inner brain" structures that are located under the cortex, in the interior of the brain, and are involved in emotion and many automatic functions such as controlling arousal and hunger).

The diagram notes the locations of the four lobes of the brain—occipital, temporal, frontal, and parietal—and the location of the Sylvian fissure, a large, highly visible crease that roughly divides the brain into top and bottom parts. Each of the lobes implements many relatively specialized systems, but for our purposes it will be most useful to group the lobes into two large processing systems: The occipital and temporal lobes are in the bottom part of the brain, and the parietal and most of the frontal lobes are in the top part of the brain. A further neuroanatomical distinction must also be made: The frontal lobe itself can be divided into a top and bottom portion, based on how these portions are connected to the parietal and temporal lobes, respectively. Thus the brain neatly divides into a top and bottom part.

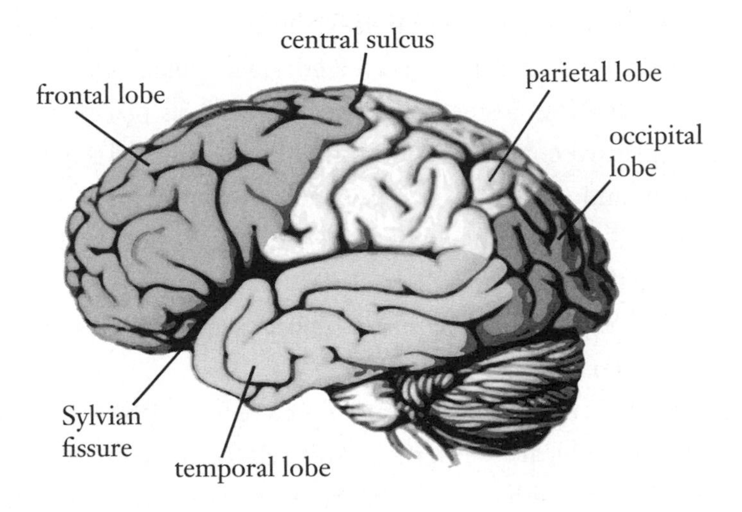

The lobes of the brain. Note that the crease along the top of the temporal lobe is the Sylvian fissure, which divides most of the bottom brain from the top brain.

The top and bottom portions of the brain have very different functions. This fact was first discovered in the context of visual perception, and it was supported in 1982 in a landmark report by National Medal of Science winner Mortimer Mishkin and Leslie G. Ungerleider, of the National Institute of Mental Health. This trailblazing study, which went largely unnoticed in the popular culture, examined rhesus monkeys. Their brains process visual information in much the same way as the human brain.

It's worth pausing to consider some details of the study by Mishkin and Ungerleider, because it provided fundamental insights into the different functions of top and bottom brain systems. The scientists trained monkeys to perform two tasks. In the first task, the monkeys had to learn to recognize which of two shapes concealed a bit of food. The shapes were three-dimensional objects (such as a striped prismatic block) that concealed small cups, one of which contained a tasty morsel. The objects were shuffled randomly each time they were presented, but the same object covered the food every time, so the animals needed to learn to recognize it in order to find the food. In the second task, both objects were identical gray placards; both placards concealed small cups, one of which contained food. Now, a small cylindrical block was placed closer to whichever placard concealed the food. The location of the cylinder was shuffled randomly each time the choice was presented, so that it was closer to one of the placards than the other—but the food was always under the placard that was closest to the cylinder. The monkeys needed to learn to recognize which placard was closest to the cylinder in order to find the food.

In short, one task required learning to recognize *shape*, whereas the other required learning to recognize relative *location.*

After each animal had mastered the two tasks, a part of its brain was surgically removed. Some animals had a portion of the bottom brain taken out (the lower part of the temporal lobe), whereas others had a portion of the top brain taken out (the rear part of the parietal lobe). The results of these operations were dramatic: The animals that had a portion of the bottom brain removed no longer could do the shape task—and could not be taught to perform it again—but they could still perform the location task well. The animals that had a portion of the top brain removed had exactly the opposite problem: They could no longer do the location task, and could not relearn how to perform it—but they could still do the shape task well.

Many later studies, including those that relied on using neuroimaging to monitor activity in the human brain while people performed tasks analogous to the ones the monkeys had performed, have led to the same conclusion: Processing in the temporal lobe (located in the bottom brain) plays a crucial role in visual recognition—the sense that we've seen an object before, that it's familiar (*I've seen that cat before*)—whereas processing in the parietal lobe (in the top brain) plays a crucial role in allowing us to register spatial relations (*One object is to the left side of the other*).

These functions occur relatively close to where neural connections deliver inputs from the eyes and ears—but processing doesn't just stop there. Rather, information about what an object is and where it is located flows to other brain areas, which do different things with that information. Researchers have shown that the top and bottom brain play specialized roles in functions as diverse as memory, attention, decision making, planning, and emotion. For example, Fraser Wilson, Séamas Scalaidhe, and Patricia Goldman-Rakic, working at Yale University, reported the results of a seminal study in 1993 in the journal *Science*. Like Mishkin and Ungerleider, they trained monkeys to perform a task, but instead of removing parts of their brains the researchers inserted tiny wires into specific areas of the brain—which allowed them to monitor the activity of individual neurons while the monkeys tried to hold briefly in memory where specific shapes were located on a screen and make simple decisions about the objects. Decisions about the locations of objects were associated with active neurons in the top brain (specifically, in an upper part of the frontal lobe), whereas decisions about objects themselves were associated with active neurons in the bottom brain (in a lower part of the frontal lobe).

As illustrated in the next figure, research examining the structure of the brain—in both monkeys and humans—has shown that the neural tracks leading from the occipital lobe do not stop in the parietal or temporal lobes, but rather continue on to the top and bottom parts of the frontal lobe, respectively. But more than that—and more important for our theory—research has documented that the top and bottom brain systems each have an internal organization that allows the various constituent brain areas to communicate and work together rapidly.

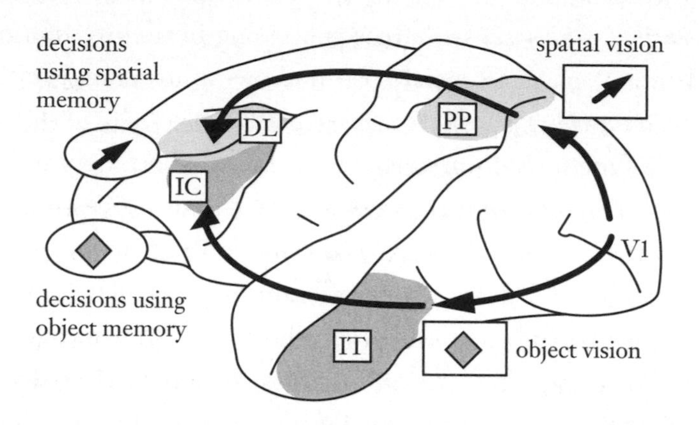

The first area to receive input from the eyes is V1 (the V for visual and the 1 to signify first). Massive neural pathways lead down to the temporal lobe (IT indicates inferior—meaning lower—temporal) and up to the parietal lobe (PP indicates posterior—rear—parietal). Both pathways continue into the frontal lobe, to the dorsolateral (upper side, DL in the diagram) and bottom parts (IC indicates inferior—lower—convexity). These pathways play a role in vision, in holding information briefly in mind to make decisions, and in other functions.[2]

Two Brain Systems

With systems in mind, let's take a quick look at what the four lobes of the brain do and how they work together in the bottom brain and top brain. We'll get into more detail later, as we continue to build the scientific foundation of our theory.

The occipital lobe (part of the bottom brain) is entirely concerned with processing visual input from the eyes, and key parts of the temporal lobe (also in the bottom brain) process auditory input from the ears. These parts of the brain organize information-bearing signals from the senses, filter noise from the signal, and then shunt the signals deeper into the brain. As Ungerleider, Mishkin, and many others have shown, other parts of the temporal lobe contain enduring memories of shapes and sounds. Hence, the signals from the eyes and ears eventually access relevant stored memories about objects, rather like a key opening the appropriate lock. These inputs from the eyes and ears are also shunted to the bottom parts of the frontal lobe, which is very tightly linked to the temporal and occipital lobes by many neural connections. Lower parts of the frontal lobe specify emotional memories.

In short, the bottom part of the brain is largely concerned with processing inputs from the senses and using them to activate the appropriate memories about relevant objects and events. For example, when you see a friend's face in a sea of strangers, you recognize her face because the input from the eyes acts like a key that unlocks the memory of your friend. Once you've activated the relevant memories, you know things about the stimulus that are not apparent in what you see—such as that she likes cappuccino, has had a lot of experience working in your industry, and often gives good advice.

Knowing what you are seeing or hearing is sometimes an end in itself (such as may occur when watching TV), but not often. Usually, we want to know what's going on around us so that we can specify goals and figure out how to achieve them. For instance, you might decide to ask your friend to get together at a favorite coffeehouse to have a cup of cappuccino—and plan to ask her advice about a problem you are having at work.

Where do such plans come from, and how are they acted on?

Devising and carrying out plans is the realm of the top-brain system. In particular, the top parts of the frontal lobe are concerned with these functions. But how does the top brain know what is being perceived? Information about where objects are located in space is so important for making plans that it is processed directly in the top brain; we need to know where objects are located in order to decide how to move them or how to move our bodies as we seek to approach or avoid them. (In our example, without such information, you couldn't have known how to thread your way through the crowd to reach and talk to your friend.) But we need to know more than just where objects are located—we also need to know what they are. Such information from the bottom brain goes to the top brain, allowing the top brain to use information about the nature of objects being perceived.

The top brain is coordinated with the bottom brain in many ways. One of the most fundamental is that while the visual and auditory perception areas (in the occipital and temporal lobes) are sending signals deep into the bottom brain, they simultaneously are sending signals up to the parietal lobe. As we saw earlier, the parietal lobe—among other things—registers where objects are located (your friend in the crowd). This information is then passed to the top parts of the frontal lobe. In fact, the parietal lobe and top parts of the frontal lobe are very tightly linked by many neural connections; these connections are so rich, and so precisely specified, that it's clear that the parietal lobe and frontal lobe work together hand in glove.

The top part of the frontal lobe also contains numerous areas that control movements. Because our movements occur in our immediate environment, to program them appropriately our brains need to know where objects are located—to reach for them, step over them, run from them, and so forth. To walk over to your friend, you need to know where she is relative to your body; to talk to her, you need to know where she is facing, and you need to position yourself close enough (but not too close!) so that she can hear you easily.

Many of the top parts of the frontal lobe stand between inputs (sensory signals) and outputs (actions), which is why these areas are involved in making decisions and devising plans. These areas do not simply receive inputs from the bottom brain about objects that we perceive, and from the parietal lobe about where those objects are located; these top portions of the frontal lobe also receive input from the bottom parts of the frontal lobe that are involved in emotion. Hence, the top parts of our frontal lobe can take into account the confluence of information about "what's out there," our emotional reactions to it, and our goals. They then play a crucial role in allowing us to formulate plans, make decisions, and direct attention in particular ways (in part by connections to the parietal lobes); they allow us to figure out what to do, given our goals and our emotional reactions to the unfolding events that surround us.

The dividing line between the frontal and parietal lobes is the central sulcus (a sulcus is a crease in the brain, and a really deep sulcus is called a fissure, as in the Sylvian fissure). Immediately in front of this roughly vertical crease, in the frontal lobe, is an area called the motor strip, which controls fine movements. And immediately behind the central sulcus, in the parietal lobe, is the somatosensory strip, which registers sensation from the body.

Why would an input and an output area be right next to each other? Here is one possible answer: Clearly, we don't always persist in a plan if it is not having the desired consequences. For instance, let's say that after you talk to your friend, you pull out your phone to make an entry in your calendar to note when you plan to meet. But the phone starts to slip out of your hand. You would quickly tighten your grip, preventing it from dropping to the ground. From a systems perspective, it makes sense to specify the feedback about how something feels in an area right next to the area that controls precise manipulations of the object: This arrangement would allow the system to use the feedback quickly, producing movements to accomplish a particular goal (not letting the phone fall, in this case).

These considerations about how different parts of the brain interact led Stephen, coauthor of this book, to organize a team to conduct a large, unprecedented analysis of the scientific literature on the functions of the top and bottom parts of the brain, as we discuss in detail in chapter 3. This analysis characterized the essential functions of the two systems:

The bottom-brain system organizes signals from the senses, simultaneously comparing what is being perceived with all the information previously stored in memory—and then uses the results of such comparisons to classify and interpret the object or event that gives rise to the input signals.

The top-brain system uses information about the surrounding environment (in combination with other sorts of information, such as emotional reactions and need for food or drink) to figure out which goals to try to achieve. It actively formulates plans, generates expectations about what should happen when a plan is executed, and then—as the plan is being carried out—compares what is happening with what was expected, adjusting the plan accordingly (for example, by adjusting your grip as the phone starts to slip from your hand).

Here's a crucial point: The two systems *always* work together. You use the top brain to decide to walk over to talk to your friend only after you know who she is (courtesy of the bottom brain). And after talking to her, you formulate another plan, to enter the date and time in your calendar, and then you need to monitor what happens (again using the bottom brain) as you try to carry out this plan (a top-brain activity). Moreover, the top-brain system prepares the bottom-brain system to classify expected objects and events, making that system work more efficiently. If you were expecting to see your friend in the crowd, this would actually be easier than noticing her without warning. The expectation (via the top brain) "primes" the recognition machinery in the bottom brain.

The systems interact in various ways, which we will discuss more thoroughly in chapter 3. For present purposes, however, the key hypothesis is that a person tends to utilize each of the two brain systems to a greater or lesser extent.

We need to emphasize that all of us use each brain system every minute of our waking lives—we couldn't function in the world without doing this. But we need to distinguish between two kinds of use: One kind is like using the brain for walking, which is largely dictated by the situation. If you see your friend and want to talk to her, you walk. The other kind is like using the brain for dancing, which is optional. You rarely, if ever, absolutely must dance. But you could learn to dance, and dancing might develop into a hobby—and you then might seize any opportunity to dance.

When we speak of differences in the degree to which a person relies on, or utilizes, the top-brain and bottom-brain systems, we are speaking of differences in this second kind of utilization, in the kind of processing that's not simply dictated by a given situation. In this sense, you can rely on one or the other brain system to a greater or lesser degree. For example, you might typically rely on your bottom brain a good deal but your top brain a little less, yielding good observations but fewer complex and detailed plans. The degree to which you tend to use each system will affect your thoughts, feelings, and behavior in profound ways. The notion that each system can be more or less highly utilized, in this sense (for reasons we examine in chapter 6), is the foundation of the Theory of Cognitive Modes.

Let's have a brief overview of this theory now.

Four Cognitive Modes

Four distinct cognitive modes emerge from how the top-brain and bottom-brain systems can interact. The degree to which each of the brain systems is used spans a continuum, ranging from highly utilized to minimally utilized. Nevertheless, for our purposes it is useful to divide the continuum into "high" and "low" categories.

	Highly Utilized Top	**Minimally Utilized Top**
Highly Utilized Bottom	Mover Mode	Perceiver Mode
Minimally Utilized Bottom	Stimulator Mode	Adaptor Mode

Mover Mode results when the top- and bottom-brain systems are both highly utilized. When people think in this mode, they are inclined to make and act on plans (using the top-brain system) and to register the consequences of doing so (using the bottom-brain system), subsequently adjusting plans on the basis of feedback. According to our theory, people who habitually rely on Mover Mode typically are most comfortable in positions that allow them to plan, act, and see the consequences of their actions.

Perceiver Mode results when the bottom-brain system is highly utilized but the top-brain system is not. When people think in this mode, they use the bottom-brain system to try to make sense of what they perceive in depth; they interpret what they experience, put it in context, and try to understand the implications. However, by definition, people who are operating in Perceiver Mode do not often initiate detailed or complex plans.

Stimulator Mode results when the top-brain system is highly utilized but the bottom-brain system is not. According to our theory, when people rely on Stimulator Mode they may be creative and original, but they do not always know when "enough is enough"—their actions can be disruptive, and they may not adjust their behavior appropriately.

Adaptor Mode results when neither the top- nor the bottom-brain system is highly utilized. People who are thinking in this mode are not caught up in initiating plans, nor are they fully focused on classifying and interpreting what they experience. Instead, our theory predicts that they are open to becoming absorbed by local events and immediate imperatives. They should tend to be action-oriented, and responsive to ongoing situations.

We will argue that each of us has a dominant mode, which is a distinctive feature of our personality—as characteristic and as central to our identity as our attitudes, beliefs, and emotional makeup. You can take a test in chapter 13 to find out which mode—Mover, Perceiver, Stimulator, Adaptor—best characterizes your dominant cognitive mode. However, our theory implies that we nevertheless sometimes adopt different modes in different contexts (as we discuss in chapter 8).

We hope that this new way of looking at the brain will give you insights into how your own brain affects your thinking, feeling, and behavior—and how that, in turn, may influence your relationships with others on the job, and in social and more intimate circumstances. Later in the book, you will find ideas—derived from the theory—that may help you decide when you should work with another person. You will find possible strategies for coping if you find yourself having to function with someone whose dominant cognitive mode rubs you the wrong way, whether at work or in a personal relationship. These ideas are not received wisdom or fully supported discoveries, but they are based on solid foundations and we believe that they are worth considering.

We've given you a lot of information, but there are just a few key points to keep in mind. To reiterate: First, the top and the bottom parts of the brain perform different sorts of tasks. Most important for our purposes, the bottom-brain system classifies and interprets sensory information from the world, and the top-brain system formulates and executes plans. Second, we also hypothesize that people variously tend to rely heavily on optional processing in the top-brain system but not the bottom one, in the bottom-brain system but not the top one, in both systems, or in neither system. Third, those four possibilities define four basic cognitive modes, which underlie how a person tends to approach situations in the world and interact with others.

The following chapters provide the scientific foundations for the Theory of Cognitive Modes, and then unpack the theory in detail. Along the way, we draw out possible implications for everyday life. We end with some thoughts about how our perspective may both help you to see yourself in a new light and help you to devise new ways to work with others and new ways to approach personal relationships. The theory may or may not be entirely correct (only time will tell), but it is rooted in solid scientific findings and is plausible—and simply thinking about it can lead to interesting insights.

Chapter 2

Roots of the Theory

The Theory of Cognitive Modes rests on a principle universally accepted today: Different regions of the brain carry out different specialized functions. In other words, functions are not spread throughout the organ. In neuroanatomical terms, highly specialized brain functions are *localized*, not *holistic*.

But this consensus about how the brain works is comparatively new, especially when the long history of brain study is considered.

Neuroscience has roots in several ancient traditions. Perhaps the oldest is the prehistoric practice of *trepanation*, which entailed opening a hole into the skull to expose the brain, apparently in an attempt to cure disease, treat trauma, or ward off demons. Another, highly developed tradition traces back to the ancient Egyptians. They practiced sophisticated medicine and surgery—but failed to grasp the true importance of the brain. They devised treatments for brain injuries, but otherwise thought so little of the organ that they disposed of it when mummifying a body for the journey to eternal life.

The great Greek philosopher and polymath Aristotle was only somewhat less dismissive, believing that the brain serves to cool the blood; he believed that the heart was the seat of intelligence and emotion—a misconception that has symbolically persisted in imagery of love and affection (try escaping it on February 14). Hippocrates, often called the "Father of Medicine," came closer to the truth. "It ought to be generally known that the source of our pleasure, merriment, laughter and amusement, as of our grief, pain, anxiety and tears, is none other than the brain," he wrote in his landmark "On the Sacred Disease," an essay about epilepsy. But Hippocrates apparently did not conceive the possibility that this three-pound mass of tissue might actually be a complex mechanism composed of multiple parts.

The concept of holistic functioning—the idea that the brain works as a single integrated organ to accomplish all of its feats, much as the liver and lungs function as integrated wholes—would endure through the Middle Ages, when the rate of progress in science in general slowed. With the Renaissance, anatomists such as the seventeenth century's Franciscus Sylvius (for whom the Sylvian fissure is named) began to push neuroscience forward.

The Oxford professor of natural philosophy Thomas Willis, a contemporary of Sylvius, was apparently the first to propose that different (large) regions of the brain were responsible for different functions. Willis's 1664 book *Cerebri Anatome*, illustrated by the great English architect Christopher Wren, proved influential in establishing the principle of localization—the idea that different parts of the brain do different things. Not everyone embraced his findings, but they received significant support some eighty years later, when the Swedish scientist Emanuel Swedenborg published a book describing neurons and areas of the brain that he believed controlled muscle movement. Foreshadowing a basic tenet of modern neuropsychology, Swedenborg also postulated that the frontal lobes give rise to critical cognitive functions. "If this portion of the cerebrum is therefore wounded," he wrote in his classic *Oeconomia Regni Animalis*,[1] "then the internal senses—imagination, memory, thought—suffer; the very will is weakened, and the power of its determination blunted."

Four decades after *Oeconomia Regni Animalis*, the Czech physiologist and anatomist Jiří Procháska took Swedenborg's observations a step further. Within larger regions of the brain, Procháska theorized, smaller parts could be identified with individual responsibility for different cognitive functions; he argued that these smaller parts, each of which he called an "organ," must act together to accomplish the complex processes of the human mind.

"It is therefore by no means improbable that each division of the intellect has its allotted organ in the brain," he wrote, "so that there is one for the perceptions, another for the will, and imagination, and memory, which act wonderfully in concert and mutually excite each other to action."

Procháska did not use the term "brain systems," but that is what he was describing (even though he characterized the systems incorrectly). He had intuited a fundamental assumption of brain function as we understand it today.

Franz Joseph Gall, Phrenologist

Ironically, it was the advent of psychology's first mass-market fad that would prove a major factor in settling the debate about whether brain function is localized or holistic—a debate that preoccupied scientists during much of the nineteenth century.

Proposed by the eccentric Viennese doctor and neuroanatomist Franz Joseph Gall, phrenology was based on the premise that the brain is the home of the mind and is divided into specialized regions. Like Procháska, Gall called these specialized regions "organs." Unlike the Czech anatomist, Gall claimed to have identified these regions and their supposed mental functions. In his uniquely titled work (translated into English), *The Anatomy and Physiology of the Nervous System in General, and of the Brain in Particular; with Observations upon the possibility of ascertaining the several Intellectual and Moral Dispositions of Man and Animal, by the configuration of their Heads*, published in 1819, Gall maintained that as a person develops, brain growth affects the structure of the skull according to the respective size and shape of each underlying organ. A bump, he asserted, signified prominence in a particular aspect of personality, whereas an indentation suggested a deficit. Thus, examining a person's skull was a way to assess the nature of his or her brain—and a way to assess the mental functions that are accomplished by specific parts of the brain. Later research showed that although skull structures indeed *do* differ from person to person, these differences do not reflect variations in the size of brain areas; cranial variations do not influence personality or mental capacity.

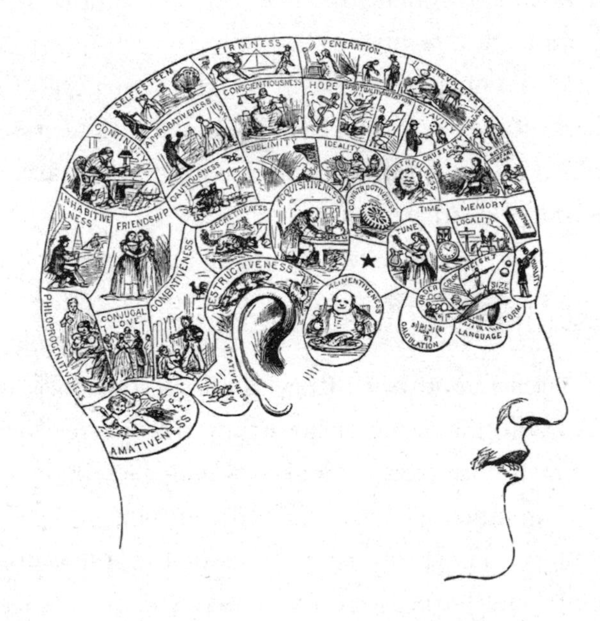

The nineteenth-century "science" of phrenology held that cognitive functions are localized in specific areas of the brain. A professional examination of a person's skull could supposedly reveal the strengths or weaknesses of each.

But in its time, Gall's theory caught fire. Here, at last, was a simple, easily understood theory of psychology. An ordinary person didn't need microscopes or fancy formulations to benefit—a painless visit to the local phrenologist did the trick. (And cost money, of course.)

The examination was straightforward: A phrenologist ran his fingertips across the scalp to discern the tangible features of the skull, and sometimes used calipers and measuring tape. He then pronounced his findings, often with the assistance of a drawing or bust of the brain's "organs." Among the original twenty-seven organs that Gall identified were those that produced *friendly attachment or fidelity; murder, carnivorousness; sense of cunning; good nature, compassion, moral sense; faculty for words, verbal memory; and theosophy, sense of God, and religion.*

A comprehensive analysis could be completed in about an hour's time. Employers hired phrenologists to screen prospective employees, and couples frequented them for insight into their relationships. Parents consulted phrenologists for advice on raising their children. Individuals seeking only to improve themselves were counseled.

Not everyone believed the claims made by the phrenologists. No less a figure than Mark Twain judged phrenology nonsense, lampooning it in *The Adventures of Huckleberry Finn* when writing of "[t]he celebrated Dr. Armand de Montalban, of Paris," a charlatan who also claimed powers of divination. Writer Ambrose Bierce was harsher, skewering phrenology in his satirical lexicon *The Devil's Dictionary* (originally titled *The Cynics' Word Book*) as "[t]he science of picking the pocket through the scalp. It consists in locating and exploiting the organ that one is a dupe with."

Phrenology never gained traction in scientific circles because certain of its fundamental assumptions are incorrect. Skull *shape* doesn't reflect cognitive functioning, and single, localized areas of the brain do not give rise to "firmness," "conjugal love," "carnivorousness," and the like. These sorts of complex functions arise from multiple cognitive and emotional processes (which are often accomplished by multiple brain areas working together).

And yet, long after Gall died, phrenology continued to attract believers and practitioners. The British Phrenological Society was not disbanded until 1967, and in October 2007 the state of Michigan extended its sales tax to include astrology, numerology, palm reading—and phrenology. A website, phrenology.org, continues to promote it today. Charts and busts are offered for sale on eBay. Some fads die hard.

Pierre Paul Broca and His Influences

Phrenology proved false, but it was not without value. It further established the concept of specifying and analyzing mental abilities; even if the mental abilities the phrenologists identified proved incorrect, the idea of breaking the mind down into components proved correct. And it reinforced an emerging consensus that brain functions are localized.

Still, some scientists in rejecting phrenology continued to reject localization. Into this continuing scientific disagreement stepped the Parisian physician and neuroanatomist Pierre Paul Broca, who in the 1860s showed that the production of speech relies on front portions of the left hemisphere—convincing evidence, he argued, for localization.

Broca reached this conclusion after studying two patients with brain damage. One, despite being cognitively functional in other respects, could say only five words (in his native French): yes, no, three, always, and "Lelo," a mispronunciation of his surname, Lelong. Broca's second, more famous, case was Leborgne: a fifty-one-year-old man who could produce only a single syllable, "Tan," whenever he attempted to speak; "Tan" became the nickname the staff gave him at the institution where he was hospitalized. Believing that these two speech-restricted men suffered brain lesions—areas of damaged brain tissue—Broca autopsied their brains after they died in 1861. As he had suspected, he found lesions in the same area of the left hemisphere in both brains. Given Broca's stature in the international scientific community, his support for localized over holistic function carried significant weight.

Broca's work influenced psychologists, including Harvard's William James, a giant in the field. Twelve years in the writing, James's landmark *The Principles of Psychology*, published in 1890, was his life's achievement. "Psychology is the Science of Mental Life, both of its phenomena and of their conditions," James wrote at the book's opening. And in a single line in the first chapter, "The Scope of Psychology," he emphatically completed the circle from brain function to thought and behavior. "Our first conclusion, then," he wrote, "is that a certain amount of brain-physiology must be presupposed or included in Psychology." The idea that consciousness itself is seated in the anatomy of the brain may seem obvious today, but in the late nineteenth century, some held it to be revolutionary.

Aided by new technologies, scientists by the 1940s not only had accepted the idea that different regions of the brain support different broad functions (such as the role of the parietal lobes in registering locations of objects in space) but also could demonstrate that smaller, identifiable areas may have very specialized functions (for instance, Area V5 plays a key role in allowing us to detect motion). A pioneer in this pursuit was the American-born Canadian neurosurgeon Wilder Penfield.

Penfield found research possibilities in epileptics who sought surgical treatment of their disease. Epilepsy occurs when the neurons in a small region of the brain begin to fire in synchrony and recruit other nearby neurons until a large portion of the brain is in spasm. This spasm creates uncontrollable movements, which are often dramatic and can be violent. In many—but not all—cases, medication can control the disease. When medication fails, patients sometimes are desperate for a treatment and volunteer to undergo new surgical procedures designed to quell or prevent the brain spasms. Penfield developed such procedures, notably the so-called Montreal procedure, in which portions of brain tissue are removed. Prior to surgery, he tried to identify the functions accomplished by specific parts of the brain, so that he did not inadvertently disrupt crucial functions. Because the brain has no pain receptors, Penfield was able to stimulate electrically the exposed brains of his patients, who had been locally anesthetized but were awake after their skulls were opened. Penfield monitored their verbal responses as he moved the site of electrical stimulation, millimeter by millimeter; in this way, he was able to map precisely small areas of the brain, showing correlations with language, motor control, and other functions.[2]

During the same period, psychologist Donald O. Hebb, a colleague of Penfield's at Montreal's McGill University, confirmed one of James's fundamental assumptions by definitively linking cognitive and biological functioning. In the introduction to his 1949 book *The Organization of Behavior: A Neuropsychological Theory*, Hebb wrote:

> "Mind" can only be regarded, for scientific purposes, as the activity of the brain. . . . Psychologist and neurophysiologist thus chart the same bay—working perhaps from opposite shores, sometimes overlapping and duplicating one another, but using some of the same fixed points and continually with the opportunity of contributing to each other's results. The problem of understanding behavior is the problem of understanding the total action of the nervous system, and *vice versa*.

And thus the stage was set. In the next chapter we return to the duplex brain, examining in more detail what's been learned about the functions of these two portions of the organ.

Chapter 3

The Duplex Brain

Franciscus Sylvius is credited with discovering the major anatomical divide between the top and bottom parts of the brain that bears his name. "Particularly noticeable is the deep fissure or hiatus which begins at the roots of the eyes (*oculorum radices*)," Sylvius wrote, in his 1663 *Disputationem Medicarum*. ". . . It runs posteriorly above the temples as far as the roots of the brain stem (*medulla radices*). . . . *It divides the cerebrum into an upper, larger part and a lower, smaller part*" (emphasis added).

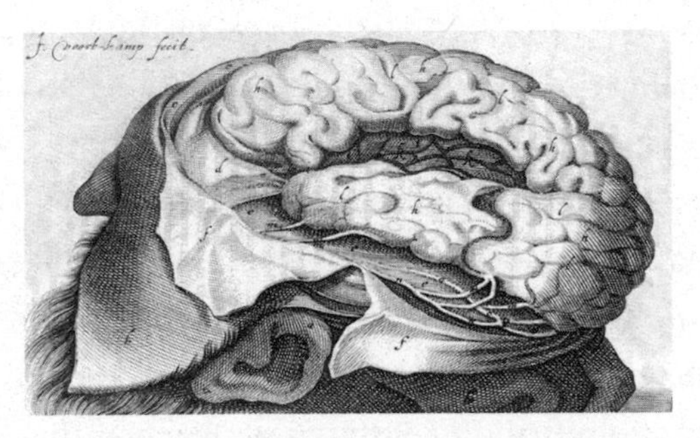

The seventeenth-century Dutch scientist Franciscus Sylvius is credited with first identifying the divide between the top and bottom parts of the brain that now bears his name. Its significance went unrecognized for centuries. *J. Voort Kamp in* Institutiones Anatomicae, *by Caspar Bartholin.*

The seventeenth-century engraving on page 27 is apparently the first illustration of the Sylvian fissure, also known as the lateral sulcus and the lateral fissure. The fissure is the dark region between the two larger areas of the folded surface of the brain (which have been pulled back to reveal the fissure).

A modern map of the brain (opposite) also highlights the all-important cerebral cortex, which is the outer covering of the brain. The cerebral cortex contains most of the cell bodies of neurons, the brain cells that process most types of information. In preserved specimens, the cerebral cortex is gray in color (hence the term "gray matter"). As we have noted, the Theory of Cognitive Modes is concerned exclusively with functions of the cortex; we are not addressing the functions of the many structures that lie beneath the cortex (and that were the focus of other theories, such as Paul MacLean's theory of the "triune brain"—which included a "lizard brain" surviving deep in the interior of the mammalian brain). Wrinkled, like the shell of a walnut, the cortex is divided into the four lobes summarized in chapter 1 (each lobe is duplicated on both sides, left and right, and thus there are a total of eight lobes: four matched pairs). Each lobe does many distinct things, and these functions typically are accomplished in small, specialized regions—usually not distributed across the entire lobe.

The occipital lobe. Located at the back of the head, the occipital lobe is the smallest of the four. As the cortical destination of most of the signals that originate in the eyes, the occipital lobe is devoted entirely to vision. Separate areas within this lobe work together to organize the varied properties of viewed objects, such as by demarcating edges of shapes and beginning to specify color. The distinctive "stars" one experiences when struck in the back of the head are the result of random firing of agitated neurons in the occipital lobe.

The temporal lobe. The temporal lobe is located under the temples, in front of the ears on the bottom of the brain; this lobe is separated from the frontal and parietal lobes by the Sylvian fissure. The temporal lobe is vital to the processing of sound and comprehension of language. It also plays a major role in entering new information into memory, some aspects of emotion, storing visual memories, defining the colors of objects, and classifying perceived objects.

As we have seen, these two lobes make up most of the bottom-brain system. This system registers sensory input, organizes it,

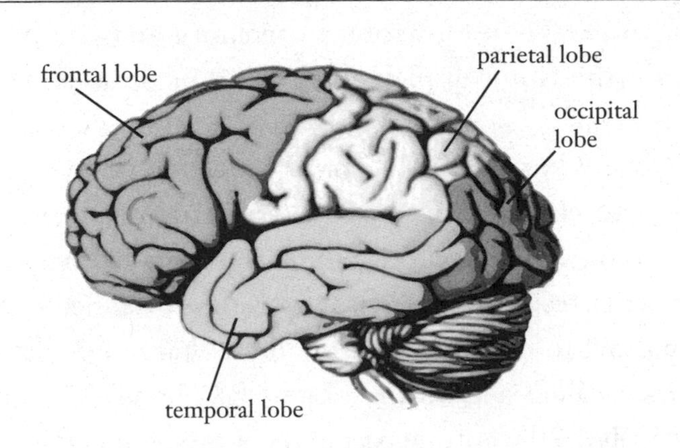

MAJOR FUNCTIONS OF THE LOBES

Frontal lobe:

- Setting up plans, making decisions, holding information briefly in mind, sequencing, directing attention, noticing disparities between what was expected and what occurred, emotional memories
- Producing speech, controlling emotional expression, guiding movements

Parietal lobe:

- Locating objects in space relative to the body, specifying relative sizes and locations of objects
- Carrying out arithmetic, touch

Occipital lobe:

- Organizing visual input into shapes, surfaces, and objects
- Sending visual information to the temporal lobe and the parietal lobe for further processing

Temporal lobe:

- Storing information in visual long-term memory; entering new information into long-term memory; hearing
- Classifying perceived stimuli, comprehending language

and uses that input to identify the object or scene.[1] When an object is identified, you can apply to it information you learned previously about similar objects. You know that apples may have the occasional worm inside; maybe you don't see one in the apple you're holding, but you found one before or learned that worms are sometimes found in apples. Similarly, when you see an old friend you know a lot about her preferences because you've previously stored this information in memory and can now access it—even if she is not displaying evidence of these preferences at the time.

The frontal lobe. Located at the front of the brain, immediately behind the forehead, this is the largest lobe in the human brain. We divide it into two parts, which reside in the top-brain and bottom-brain systems, respectively. These two parts are defined by how they are connected to other parts of the brain. Large bundles of nerve fibers, each called a fasciculus (Latin for "bundle"), connect the temporal lobe to the bottom part of the frontal lobe, which is part of the bottom-brain system. This part of the frontal lobe is particularly involved in emotion and reward. In addition, another fasciculus connects the parietal lobe to the top portions of the frontal lobe, which is part of the top-brain system. These top portions of the frontal lobe are critically involved in controlling movements, producing speech, and guiding the search for specific memories, as well as in some aspects of directing attention (both to the outside world and to internal events) and in reasoning, making decisions, short-term memory, and "executive function"—which include the capacity to formulate plans and to anticipate consequences resulting from actions. The top inside portion of the frontal lobes contains the anterior cingulate cortex, which plays a role in comparing what was expected to occur with what is perceived to have occurred, and registering disparities between the two. Finally, at the rear of the top portion of the frontal lobe lies the "motor strip," which controls fine movements of specific parts of the body.

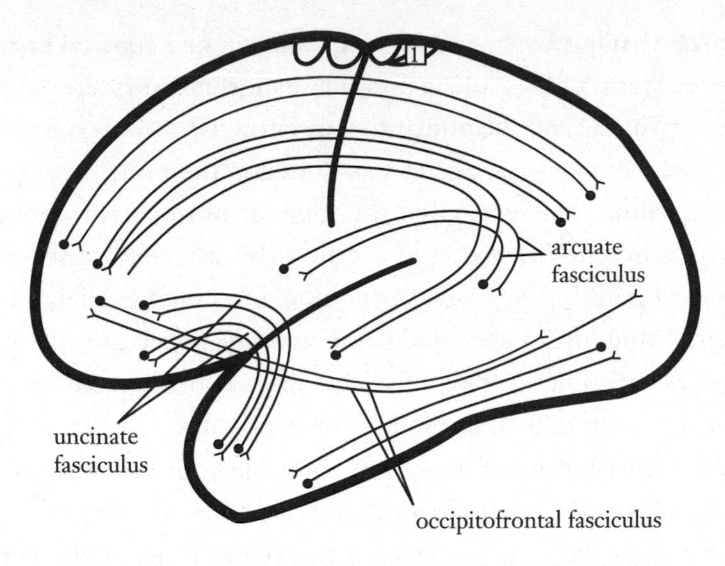

The major long-range connections (fasciculi) in the brain. Note that these connections define which parts of the frontal lobe are in the top-brain system and which parts are in the bottom-brain system.

We must note what, at first blush, appears to be an important exception to our generalization about what the top brain does versus what the bottom brain does: Broca's area, which is involved in speech production, is located in what we would call the bottom-brain portion of the frontal lobe. This area receives rich connections from the top parts of the frontal lobes, from the temporal lobe, and from motor, somatosensory, and parietal regions. The pattern of connectivity suggests that it functions in part as if it were in the top brain, as we would expect if it controls the mouth, tongue, lips, and vocal cords during speech production. However, the area has also been shown to play an important role in language comprehension—as we would expect if it were performing bottom-brain functions. Moreover, this area is activated when people try to interpret the meanings of other people's actions—again, a bottom-brain sort of function. Consistent with these functions, this area has also been implicated as playing a role in allowing people to understand the meaning of nonverbal gestures.[2] Thus, Broca's area appears to play a role in classifying and interpreting input, as we would expect from its anatomical location. However, it also plays a role in producing movements of the vocal apparatus—which again demonstrates that the two brain systems, top and bottom, constantly interact and work together.

The parietal lobe. Located at the top rear of the brain, above the occipital lobe, the parietal lobe is crucial for a variety of functions, not just registering spatial information (as noted earlier). These functions range from specifying orientation in space to the ability to perform arithmetic. This lobe also plays a crucial role in attention, both to external affairs and to internal states. It appears to play a role in consciousness—but so do all the other lobes. In addition, sensations from different parts of the body are registered in specific sections along the somatosensory strip, at the front of the parietal lobe—with sensations from the left half of the body feeding into the right parietal lobe, and sensations from the right half of the body feeding into the left parietal lobe.

The top-brain system thus comprises the parietal lobe and the top portion of the frontal lobe. Getting a bit more technical: The *top brain* consists of the entire parietal lobe and what is known as the dorsolateral prefrontal cortex—*dorso* meaning "back" in four-legged animals (which corresponds to "top" for upright humans); *lateral* meaning "side"; and *pre-* meaning the parts of the frontal lobe that are in front of the motor areas, which are in the rear portions of the frontal lobe. This dorsolateral prefrontal cortex is the top (and larger) portion of the frontal lobe. The top brain also includes the motor areas and the medial (middle) frontal cortex, notably the anterior cingulate cortex.

The *bottom brain* consists of the occipital and temporal lobes, the lower parts of the lateral frontal cortex, the orbitofrontal cortex (*orbito* because it is located right over the orbits in which the eyes reside; the orbitofrontal cortex is also called the ventromedial frontal cortex, and plays a special role in memories of emotional events), and the frontal pole (at the very front of the lobe, which is involved in integrating various forms of information).

Astonishing Ideas

Shortly after Ungerleider and Mishkin's landmark paper was published in 1982, Stephen went to their lab in Bethesda, Maryland, to meet them and see firsthand some of their experimental animals and methods. Mishkin discussed their finding that the bottom parts of a monkey's brain are involved in recognizing shapes but not in specifying spatial locations, with the reverse holding true for the top parts, which are involved in specifying locations but not in recognizing shapes. At the time, Stephen found this idea astonishing: To his knowledge, every theory in psychology and every artificial intelligence (AI) program designed to produce computer vision had ignored this distinction. (Many people tend to think of AI as a more contemporary field, but it in fact became a going concern in the mid-1950s.) During that period, AI had a larger influence on cognitive psychology than did studies of the brain; cognitive psychology focuses on the scientific study of mental processes. Stephen had dabbled a little in AI and was very familiar with the field—but was not very familiar with neuroscience.

Look through your own eyes at the space around you: Doesn't it seem irrefutable that shapes are firmly rooted in particular locations? Logically, how could it be otherwise? This compelling sense of each-object-in-its-place might seem to imply that the same part of the brain would be involved in processing both shape and location.

But as Ungerleider and Mishkin had demonstrated, the science is clear: Our impressions are misleading. The two functions are separated in the brain.[3] The fact that separate systems are used to process *what* an object is versus *where* it is implies that the two sorts of information must be "glued together" at some later phase of processing—and that idea in turn suggests that errors can occur. Intuitively, it may seem weird that you could recognize an object but nevertheless fail to see it in the correct location, but the scientific literature has clearly shown that such errors do indeed take place and can occur surprisingly frequently when objects are viewed only briefly. In a now classic experiment,[4] Anne Treisman and Hilary Schmidt showed that when people saw sets of colored shapes (for example, a square, a triangle) very briefly, they later made "illusory conjunctions" in almost one-fifth of the trials: They mentally combined features from objects that were in fact shown in separate locations. For instance, if they had seen a small red square in one location and a large blue circle in another, they might report having seen a small blue square—mistakenly combining features from objects that had appeared in different locations.

In short, the counterintuitive idea that shape and location are broken apart in the brain has received solid scientific support: The bottom-brain system processes information about the properties of an object (such as its shape and color) separately from the top-brain system, which processes information about the object's location. And the outputs from the two systems are not always combined properly.

Given the existence of the Sylvian fissure, it was only natural for a scientist to wonder: *Why* is the brain organized into these anatomically distinct top and bottom systems? Why not just one large, single system? And a more practical question: What insights might we gain into human psychology by exploring the implications of what Ungerleider and Mishkin had shown with their monkeys?

Further research with animals and people in labs around the world began to uncover more and more evidence of the importance of the top/bottom division. Stephen's own efforts at unraveling this puzzle began in earnest in the late 1980s, when he and his colleagues created a computer program that could both recognize simple shapes and indicate where they are, using simulated neural networks that corresponded to the top- and bottom-brain systems.

Such simulated neural networks have sets of "nodes," which

roughly correspond to individual neurons. The nodes may be organized in various ways. Typically, one set (so-called input nodes) is connected to receive stimulation from outside the network, one set (output nodes) sends information from the network to the outside, and a third, intermediate set falls between these two groups. The nodes can be connected up in different ways, which change the outputs that particular inputs produce.

In these computer models, the top and bottom systems corresponded to different groups of the third type of nodes and their connections to input and output nodes. In some versions of the models, the top- and bottom-brain systems were entirely separate, and in other versions they shared some of the same intermediate nodes.

Logically, it makes sense to think of the two systems as separate: The bottom system needs to recognize objects no matter where they appear, and hence throws away information about location—whereas the top system depends on the very information that is discarded by the bottom. In fact, experiments with the computer models convincingly demonstrated that the simulated neural networks operated best when they could use a clear divide-and-conquer strategy, having a distinct bottom-brain system that ignores information about location in order to recognize shapes no matter where they appear but also having a distinct top-brain system that relies on such information. This sort of division of labor was much more effective than using a single network to do both jobs or even having two partially overlapping networks.[5] Clearly, the two functions are different: The system operated much better when it was organized into distinct top and bottom parts.

Four years later, Stephen and his colleagues studied one way that the two brain systems can interact, asking whether the top-brain system could, in some situations, fill in for the bottom-brain system. Turning from computer models to people, they studied a patient who had suffered a stroke. Blood had been prevented from reaching the bottom part of the brain (probably because a blood clot had blocked an artery), and the neurons in a portion of the bottom-brain system had died.

Because this patient's bottom-brain system was impaired but his top-brain system was largely spared, the researchers hypothesized that he would be forced to use top-brain processes in situations in which the bottom-brain system is ordinarily used. Their hypothesis proved to be correct. The patient required more time than usual to recognize patterns that had more distinct locations in them, because he was registering each location separately instead of seeing an overall configuration—as he would have done prior to the damage and as was done by normal control participants. By analogy, instead of seeing a face as a single pattern, he was registering separately the locations of the eyes, the nose, and the mouth. Unlike people who had not suffered this sort of brain damage, and who recognize objects as a single pattern by using the bottom-brain system, this patient seemed able to use the bottom-brain system only to register individual parts and had to use the top-brain system to note each part's location separately, and painstakingly piece together what he was seeing.

The distinction that Ungerleider and Mishkin had made, between "what" (for the bottom system) and "where" (for the top), seemed to hold up well in many different studies of human patients who had brain damage. But another interpretation of the functions of the top and bottom brain systems was gaining ground: Researchers Melvyn Goodale and A. David Milner, writing in *Trends in Neurosciences* in 1992,[6] proposed an alternative theory that did not posit that the top-brain system registers location. They had studied a patient who apparently had rather diffuse brain damage that disrupted specific aspects of both brain systems, and discovered that this patient had difficulty consciously recognizing shapes, sizes, and even the orientation of a single line—but had little difficulty in using such information to guide movements. For instance, she could effortlessly orient a card so that it fitted into a slot that was tilted to different degrees. A series of detailed studies of this patient convinced Goodale and Milner that the proper distinction should be not between what and where but rather between what and how—the how being how one guides movements.

However, other researchers soon showed that patients who had damaged parietal lobes had trouble registering spatial relations even when this information had nothing to do with guiding movements; for example, they might have trouble noting that one object was to the left of another.[7]

It's not that Goodale and Milner were wrong—the top-brain system *does* in fact play a crucial role in controlling movements—but rather that their view was too limited; the top-brain system not only controls movements but also does lots of other things (including specifying where objects are located). The top system is crucial in specifying spatial information more generally, not simply when it is used to guide movements.

The limitations of Goodale and Milner's views underscore a fundamental problem with both the what/where and the what/how distinctions, and the problem is similar to the one that besets the left brain/right brain story: Both rely on dichotomies. The brain is just too complicated for any simple, binary distinction to get us very far when we try to characterize what large regions of the brain do. A simple black-or-white explanation is, well, simplistic. Rather, the brain processes information—and any tenable characterization of its function must specify a system that processes information.

Meta-Analyses

The only major attempt to characterize multiple aspects of information processing in the top-brain and bottom-brain systems was reported by Grégoire Borst, William Thompson, and Stephen, in an extensive 2011 analysis of the neuroscientific literature published in *American Psychologist.*[8] The team conducted a "meta-analysis," which reviewed the results of many previous studies, looking for patterns that cut across them. This meta-analysis required the researchers first to scour the scientific literature, looking for three types of studies that specifically implicated top- versus bottom-brain function: those investigating the effects of damage to the top versus bottom brain on human cognition and behavior, those in which the top or bottom brain was activated (as detected by neuroimaging) while people performed particular tasks, and those that used a technique called *transcranial magnetic stimulation* (TMS) to disrupt top or bottom brain activity temporarily in specific locations. TMS relies on delivering strong magnetic pulses to a particular region of the brain, which temporarily impair neural functioning in that region. The researchers found over one hundred pertinent published studies.

The results of this meta-analysis were clear: Four characteristics predicted when top-brain areas are involved, namely: (1) The task does not require shape processing; (2) spatial relations must be used (such as those needed to decide which of two objects is closer to a third object—so specifying "where" is indeed part of what the top-brain system does, as Ungerleider and Mishkin claimed); (3) sequencing is required (of the sort involved in carrying out plans); and (4) movement must be detected. And three characteristics predicted when bottom-brain areas are involved: (1) Parallel processing is used (this occurs when two or more processes are running simultaneously, such as when an object must be compared with many stored objects at the same time); (2) spatial relations are not required; and (3) sequencing is not required.

But this first meta-analysis did not find that the bottom-brain system was implicated when participants had to classify stimuli. (This need arises when people must assign stimuli to categories, such as when people name objects.) Why not? A closer look at the studies immediately suggested a hypothesis: Some of the studies involved classifying not shapes (a function that relies on the bottom-brain system) but rather *spatial* information, such as when one is deciding whether an object is to the left or to the right of another (a top-brain system task). When the meta-analysis was conducted again without including the results from studies that required participants to classify spatial information, the expected result emerged: The bottom-brain system is clearly involved in classification of properties of objects (so specifying "what" an object is does indeed occur in the bottom-brain system, as Ungerleider and Mishkin previously claimed).

Nonetheless, it was theoretically possible that the differences found between the top- and bottom-brain systems reflected another sort of brain divide. Perhaps the left/right distinction *did* underlie the more fine-grained analysis done in this study. Or, maybe front/back was the critical division. It was easy to evaluate these possibilities statistically, and the results of further analyses were clear: Even after possible effects of both left/right and front/back differences were statistically controlled, the top-brain areas were found to mediate processing of spatial relations and sequences as well as movement, whereas the bottom-brain areas were found to mediate classification.

In short, the meta-analysis confirmed the importance of the top/bottom distinction.

In spite of the fact that the top-brain/bottom-brain distinction does not emerge solely from processes in the left versus the right brain or in the front versus the back brain, this does not imply that the different portions of the top and bottom systems have identical functions.

To explore this possibility, the research team examined the top-front, top-back, bottom-front, and bottom-back parts of the brain. These additional analyses showed that the top-front part of the brain plays a special role in mediating sequencing, whereas the top-back plays a special role in processing spatial relations and movement; in contrast, the bottom-back brain is particularly involved in parallel processing (running more than one process at the same time). These findings were important because they mirror previous findings that have been reported in the scientific literature, which show that the top parts of the frontal lobe are particularly important for sequencing and the parietal lobe is particularly important for coding spatial relations and movement. But, crucially, these two portions of the top-brain system typically work together; they are tightly bound, both because of their rich anatomical connections and because of how they function. As we noted earlier, many—if not most—plans (which necessarily involve setting up sequences of actions) require us to move about in space (such as occurs when you decide to get up from your chair and walk to the next room, going through doors and not hitting furniture), and hence we use spatial information when setting up and carrying out such plans.

What about the what/where and what/how distinctions? That is, did these distinctions account for patterns of findings in the literature? When the data were analyzed to investigate the usefulness of these distinctions, both dichotomies fell short: They failed to organize the entire set of data anywhere near as well as the way we have characterized the top- and bottom-brain systems in this book. This meta-analysis, the only one of its kind to date, provided strong evidence for the view of top-brain and bottom-brain functioning that we rely on in this book.

One of the reasons we are interested in the top brain/bottom brain distinction rests on the assumption that people differ in how much they utilize the two systems. So far we've said little about such individual differences. In the next chapter, we focus on one way that the two systems are used—and show that people do indeed differ in this domain.

Chapter 4

Reasoning Systems

Please indulge us by answering the following "trick" questions from memory:

In what hand does the Statue of Liberty hold the torch?

Is the angle formed by the hands of a clock at 3:05 larger than the angle formed at 8:20?

What shape are Mickey Mouse's ears?

Which is darker green: iceberg lettuce or spinach?

We are really interested not in the answers but rather in how you went about trying to arrive at them. (But in case you are interested, the answers are: right, no, circular, spinach.) The vast majority of people report that they visualize the objects when trying to answer these questions. That is, they use *mental imagery*.

Mental imagery is in many ways like perception, which occurs when, for example, you are looking at a car and registering its size, shape, and speed (or, for other sensory modalities, when you are listening to a song, feeling the wind in your hair, or smelling a flower, or otherwise are aware of what your senses are registering). The difference with mental imagery is that the stimulation is not coming directly from the sensory organs—the eyes, the ears, and so forth—but instead arises when you access information you've previously stored in memory. You couldn't address these four questions if you hadn't seen those objects before and hadn't been able to access stored memories of them. Mental imagery gives rise to the experience of "seeing with the mind's eye," "hearing with the mind's ear," and the like—but mental images are not like *afterimages*, which are images that remain briefly after the senses are activated (for example, as will occur after you look at oncoming headlights at night, and "see" them lingering afterward when you gaze to the side). Unlike afterimages, mental images can be created at will, and you can often retain them for at least a few seconds.

The four questions we posed are trick questions not simply because they require visual mental imagery. The salient point here is that the top- and bottom-brain systems played different roles when you answered the first two questions compared to when you answered the second two questions.[1]

In the past two decades, neuroimaging has played a large role in the study of mental imagery. This modern technology allows researchers to identify which parts of the brain are most active at a given point in time, and it usually works by tracking how much blood has been pulled into an area. Blood flow increases to meet the increased metabolic needs of more active tissues. Neuroimaging has allowed scientists to demonstrate that mental images that involve spatial relations—such as those evoked in answering the questions about the Statue of Liberty and the clock—largely activate the top-brain system. Just as the top-brain system processes spatial relations during perception, this system processes spatial relations during mental imagery. And neuroimaging has shown that mental images of shapes and colors—such as those evoked in answering the questions about Mickey Mouse and leafy greens—largely activate the bottom-brain system. Just as the bottom-brain system processes shape and color during perception, this system processes shape and color during mental imagery.

Evidence that the top- and bottom-brain systems play a causal role in such processing also comes from studies of people who have sustained brain injuries. Patients who have suffered damage to the top-brain system have difficulty answering questions like the first two—but do fine when answering questions like the last two. Patients who have had damage to the bottom-brain system have exactly the reverse profile: They are able to answer questions about spatial relations but have difficulty answering questions about shape or color. These effects were strikingly demonstrated in a 1985 study conducted at the Massachusetts General Hospital by David Levine, Joshua Warach, and Martha Farah.[2] In this study, patients with damage to the bottom-brain system were found to have great difficulty when asked to describe the difference in appearance of a bear versus a lion, a determination requiring a grasp of shape and color—but they did not have much trouble describing what direction Chicago is from Boston (these patients lived in Boston), a determination requiring a grasp of spatial relations. Patients with damage to the top-brain system had exactly the opposite problem: They could readily describe the appearance of animals but were stymied when asked to determine direction.

Bringing this back to Ungerleider and Mishkin's landmark 1982 study with monkeys, damage to the bottom brain disrupted the ability to use mental imagery to assess "what" something is, whereas damage to the top brain disrupted the ability to use mental imagery to assess "where" it is.

Think Different

A lot has been learned about individual differences in mental imagery in the top- and bottom-brain systems. And most of this knowledge has emerged from an impressive series of studies conducted by a group headed by Maria Kozhevnikov, now at the National University of Singapore and the Massachusetts General Hospital in Boston.[3] These researchers have shown that people differ in how easily they can use the two types of imagery, the "what" and the "where"—and, crucially, that these differences are related to important aspects of daily life.

This series of studies was one of the inspirations for our Theory of Cognitive Modes. The researchers demonstrated a crucial link between top-brain and bottom-brain anatomy and function.

The first study in the series, published in 2002,[4] addressed a conundrum: For years, some researchers had maintained that certain people are "visualizers" and others are "verbalizers"—but, in spite of many efforts, the evidence for this distinction was very weak. That is, these earlier researchers believed that "visualizers" rely on visual mental imagery when they think (for example, mentally picturing likely outcomes of planned events) whereas "verbalizers" prefer to use language (for example, reasoning using logical sequences, such as, "If it is going to rain, I should bring an umbrella"). Various questionnaires had been developed to assess such differences, with a visualizer-verbalizer questionnaire requiring the test-taker to respond True or False to each of a set of statements, such as "I enjoy doing work that requires the use of words" (an item that verbalizers endorse) and "My daydreams are sometimes so vivid, I feel as though I actually experience the scene" (an item that visualizers endorse).

As intuitively appealing as this distinction is to many, questionnaires designed to assess visualizer versus verbalizer preferences generally have not predicted how people learn, think, or behave. After many years of research, the bottom line is that this distinction does not predict behavior well.

Kozhevnikov had a key insight into the problem with this distinction and with the questionnaires inspired by it: "Visual" is too coarse a category—as we've discussed, spatial information is processed separately from information about properties of objects, such as shape and color. Kozhevnikov (and her then-colleagues Mary Hegarty and Richard E. Mayer, at the University of California,

Santa Barbara) examined the relation between how people scored on tests of visualization versus how they scored on tests of spatial abilities. One of the tests of spatial abilities was the Paper Folding Test, a standard measure already used by other researchers. In this test, people are required to look at a series of drawings of a square sheet of paper, which is folded, and then folded again (two or three times). The last drawing in the series shows where a hole is punched all the way through the folded sheet. The test-takers then must examine five drawings of an unfolded sheet, which show different locations where the hole would appear, and indicate which one illustrates how the unfolded paper would actually look.

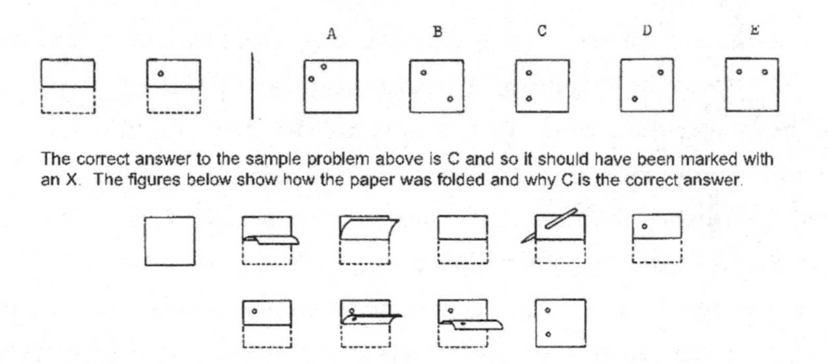

An example of an item in the Paper Folding Test, with an explanation of the correct answer. Top row: In this test a person is shown a sheet of paper that is folded in a particular way, and then a hole is punched through the folded sheets (on the left). The participant is asked to select which of the unfolded alternatives (showing where the holes occurred, on the right) is correct. Bottom: A visual explanation of why C is the correct answer. *With kind permission from Springer Science+Business Media:* Memory & Cognition, *"Spatial Versus Object Visualizers: A New Characterization of Visual Cognitive Style," Vol. 33, issue 4, January 1, 2005, Maria Kozhevnikov.*

The crucial finding reported by Kozhevnikov, Hegarty, and Mayer was that visualizers scored either very high *or* very low on the spatial abilities tests—not in the average range. For comparison, verbalizers had the usual pattern found with most tests, where the majority of people score in the average range on the spatial abilities tests, with only a few scoring either very high or very low.

In short, these researchers found that people who score as "visualizers" can be organized into two types: those with high spatial ability and those with low spatial ability. The Theory of Cognitive Modes developed in this book grew in part out of this discovery, but the new theory benefited from subsequent findings that further articulated this idea and connected it to the top-brain and bottom-brain systems.

Imagining Brain Systems

A crucial step was to make the connection from visualizer abilities to the top- and bottom-brain systems—and that is exactly what Kozhevnikov, then working at Harvard with Stephen and Jennifer Shephard, reported in a collection of studies that appeared in an article published in 2005 in the journal *Memory and Cognition.*[5] The key idea from this team was that one group of visualizers is adept at using the top-brain system to perform spatial operations in mental imagery, whereas another group is adept at using the bottom-brain system to generate and "inspect" high-resolution mental images of objects.

In the first study in this report, the researchers gave the Paper Folding Test, which requires participants (college students, in these studies) to use spatial imagery, relying in large part on the top-brain system. They also administered a test of mental image vividness, which required participants to visualize objects and report how vivid the mental image appeared (rating it on a five-point scale); this sort of task requires bottom-brain, object imagery. And they asked the participants to fill out a traditional visualizer-verbalizer questionnaire, which requires the test-taker to respond True or False to each of a set of statements (as noted earlier).

The results were striking: As they expected, the visualizers could be divided into two categories, those who performed well with spatial imagery tasks and those who performed poorly. Moreover, the visualizers who were poor at spatial imagery actually rated their images as more vivid than did the visualizers who were good at spatial imagery. This finding suggested that those who were poor at spatial ("where") imagery were good at object ("what") imagery, and vice versa.

The Harvard team followed up these findings by using a set of objective tests, in a second study included in the 2005 *Memory and Cognition* paper. Specifically, in order to assess how well people can use the top-brain system in spatial imagery, they asked a group of new participants to perform two tasks.

In the *mental rotation task*, participants were asked to compare a pair of three-dimensional-appearing angular figures that were rotated in different ways and decide whether the two figures had the same shape regardless of how they were rotated. People typically require more time when one of the two figures must be "mentally rotated" a greater amount to line up with the other figure in the pair. To get a sense of this process, visualize the uppercase version of the letter "n," and imagine that it rotates 90 degrees clockwise—is it now another letter? If so, what is that letter? (Yes, and it's the letter "Z.")

In the *embedded figures task*, participants were asked to decide whether each of a series of visualized pictures embodied specific spatial characteristics; the relevant questions focused on whether it had a certain specific embedded characteristic, such as lines that formed a T shape, which involves locating the spatial relations among lines.

In contrast, to assess how well people can use bottom-brain mental images of shapes, Kozhevnikov, Kosslyn, and Shephard created two new tasks to be used in this study: grain resolution and degraded pictures.

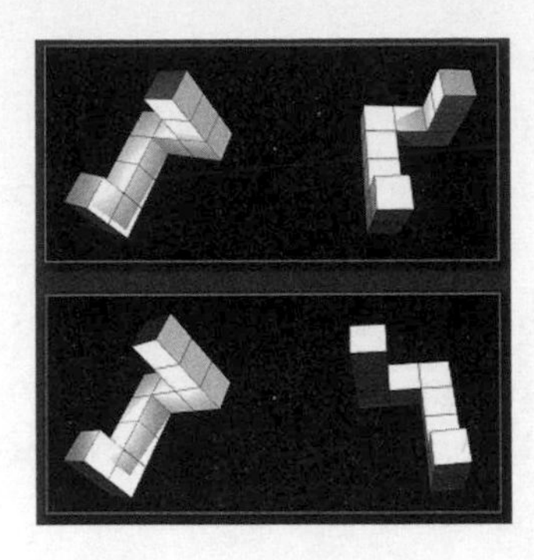

An example of a stimulus used in the mental rotation task. Participants were asked to mentally rotate one of the objects in each pair so that it lines up with the other, and then to compare the two objects to decide whether their shapes are identical or whether one is a mirror image of the other. *With kind permission from Springer Science+Business Media:* Psychonomic Bulletin & Review, *"Training Generalized Spatial Skills," Vol. 15, no. 4, January 1, 2008, Rebecca Wright.*

In the *grain resolution task*, the participants were asked to visualize pairs of named objects and decide which one has a surface with "finer texture or denser grain." ("Denser grain" meant more dimples or bumps per inch on the surface.) For example, mentally compare a strawberry and a blueberry—which one has finer texture or denser grain? (Answer: The blueberry has finer texture.) Or how about the surface of an (unpeeled) orange versus a golf ball? (Answer: The orange has the finer texture.)

In the *degraded pictures task*, people tried to name the object shown in a line drawing that was missing random segments and was overlaid with random line segments; when performing this task, people tend to look for patterns that suggest an object, and then visualize the entire object and see whether it fits the remaining line segments.

In both of these tasks, the researchers measured how accurate the participants were and how long they took to respond.

The results were just as expected:

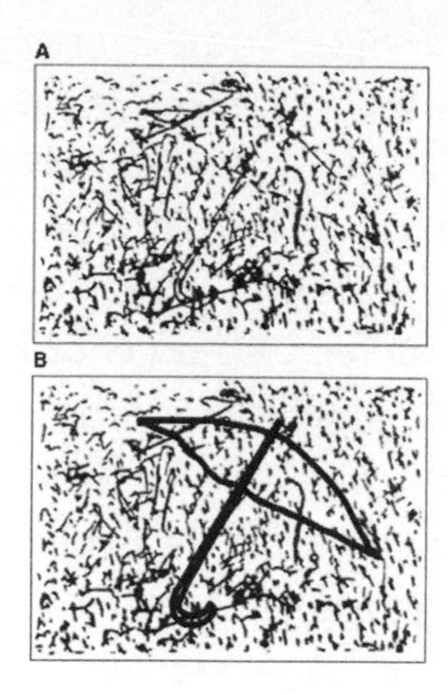

Top: An example of a stimulus used in the degraded pictures task. The participants are asked to name the object, in spite of the interference from the random line fragments placed over the drawing of it. Bottom: The outline of the embedded drawing (shown here to illustrate it; participants never saw this drawing during the test). *With kind permission from Springer Science+Business Media:* Memory & Cognition, *"Spatial Versus Object Visualizers: A New Characterization of Visual Cognitive Style," Vol. 33, issue 4, January 1, 2005, Maria Kozhevnikov.*

On the mental rotation and embedded shapes tasks, which rely critically on the top-brain system, participants who scored as high-spatial visualizers (having scored as "visualizers" and also having high spatial ability, as in the original study Kozhevnikov and her colleagues reported, summarized above) performed better than participants who scored as low-spatial visualizers. This makes sense because the tasks require proficiency in spatial imagery.

On the grain resolution and degraded pictures tasks, which rely critically on the bottom-brain system, participants who scored as low-spatial visualizers performed better than participants who scored as high-spatial visualizers. This makes sense if these people were good at using their bottom brains to visualize shapes.

The conclusions:

Spatial and object mental imagery are different, and the reason why is easy to formulate: Spatial mental imagery relies especially on the top-brain system, and object mental imagery relies especially on the bottom-brain system. Moreover, people can be good at one type of imagery while being bad at the other type. This finding was critical in the evolution of the Theory of Cognitive Modes.

One more study from the 2005 paper by Kozhevnikov, Kosslyn, and Shephard is worth noting. It posed the question: Does the distinction between the two sorts of imagery matter?

In this study, ten professional visual artists (painters, photographers, and interior designers) and fourteen scientists (physicists and engineers) were given the Paper Folding Test (assessing top-brain functioning) and the grain resolution test (assessing bottom-brain functioning). They were also shown a graph that charted the position of an object over time. The results were intriguing: The scientists performed the top-brain, spatial task better than the artists, but the artists performed the bottom-brain, grain resolution task better than the scientists.

But more than that, the two groups differed in how they interpreted the graph: The scientists tended to see it as an abstract representation of changes in position over time. By contrast, the artists understood the graph "as a literal pictorial illustration of a situation or as the path of the actual motion and did not attempt to interpret the graph as an abstract schematic representation," the paper reported.[6]

The artists and scientists clearly visualized differently. Why? Kozhevnikov, Kosslyn, and Shephard suggested that "[d]ifferent professions might promote object versus spatial imagery, or perhaps people with one sort of imagery select a field on the basis of their imagery abilities or preferences." These two alternatives represent the sort of question that lies at the heart of the nature/nurture debate.

From Harvard, Kozhevnikov moved to Rutgers University, and she and her new colleagues took the next step, devising a way to measure whether individual people preferred to use top-brain-based, spatial imagery or bottom-brain-based, object imagery. The test that Kozhevnikov, together with Olesya Blazhenkova and Michael A. Motes, reported in 2006 is called the Object and Spatial Imagery Questionnaire.[7] This test characterizes the two types of visualizers.

"For instance," they wrote, "object imagers prefer to construct colourful, high-resolution, picture-like images of individual objects and to encode and process images holistically, whereas spatial imagers prefer to construct schematic representations of objects and spatial relations among objects, generate and process images part by part, and are capable of performing complex spatial transformations."[8] Test-takers rate the degree to which they agree or disagree with statements. Some of these items implicate object imagery, such as "My mental pictures are very detailed precise representations of the real things" and "I can close my eyes and easily picture a scene that I have experienced," and some implicate spatial imagery, such as "I can easily rotate three-dimensional geometric figures" and "I prefer schematic diagrams and sketches when reading a textbook." The test is scored to indicate the degree to which the test-taker has a predisposition to use object imagery and/or spatial imagery.

To develop this test, these researchers formulated a set of such items and asked a group of over two hundred college students (selected locally) to indicate (using a five-point rating scale) the extent to which they agreed or disagreed with each item. After the results were in hand, the researchers used factor analysis (a mathematical technique, which we will discuss later when we describe how we developed the new test of cognitive modes) to determine which items were most strongly related to a spatial versus an object factor. Based on these analyses, they selected fifteen items of each type, which became the final version of their test. Blazhenkova, Kozhevnikov, and Motes validated their new test in several ways. For example, they showed that scores on the items that assess spatial imagery, and not scores on the items that assess object imagery, predicted how well people perform spatial tasks (such as the Paper Folding Test and mental rotation)—and scores on the items that assess object imagery, and not scores on the items that assess spatial imagery, predicted how well people performed the degraded pictures task and how vividly they rated their visual mental images.

In addition, as part of the process of validating the test, Blazhenkova, Kozhevnikov, and Motes gave it to members of different professions, which included visual artists (such as painters and designers), scientists (such as computer scientists, physicists, and biologists), and humanists (such as historians, philosophers, and literature professors). As expected, they found that visual artists scored higher on object imagery than did scientists or humanists, whereas scientists scored higher on spatial imagery than did artists or humanists. This pattern was predicted by analyzing what sort of imagery was most commonly used in the different professions.

In a subsequent study, published in 2010 in the journal *Cognition*,[9] Olesya Blazhenkova and Kozhevnikov provided more evidence that members of different professions tend to use different sorts of imagery. These researchers also made an important observation that is crucial for the new test we present in chapter 13: The fact that scores on tests requiring people to reflect on their preferred type of imagery predict actual performance "indicates that individuals are usually aware of their most efficient mode of visual information processing, and that self-report measures could be reliably used to identify an individual's particular strengths and weaknesses in the use of object or spatial modes of information processing."[10] Although you cannot directly monitor how much you rely on your top-brain and bottom-brain systems, you can monitor the consequences of using these systems: You can observe your own behavior—and on this basis we can ask you to evaluate test items, and then can draw inferences about your dominant cognitive mode. Of course, it is not news that people can often reflect on their preferences and that such reflections can predict their behavior. In fact, many such tests have been validated and have been shown to predict behavior—although this is not always the case, as Daniel Kahneman so convincingly documents in his brilliant book *Thinking, Fast and Slow*.[11] Thus, it is reassuring to know that self-report scores on tests like the cognitive modes self-assessment you will find in chapter 13 (which was developed in much the same way as the imagery test) have in fact been shown to predict behavior.

Resolving an Apparent Contradiction

Before continuing, we must address an apparent contradiction between findings just summarized and a key idea behind our Theory of Cognitive Modes: On the one hand, the results from the new self-assessment test in chapter 13 reveal that scores reflecting the use of the two systems are very weakly, if at all, statistically related. This means that people could be good at both types of processing, bad at both types of processing, or good at one but bad at the other. On the other hand, the experimental studies of mental imagery summarized above suggest that people who are relatively good at top-brain, spatial imagery are relatively bad at bottom-brain, object imagery—and vice versa. It's important to examine this apparent contradiction because the very idea of the four modes rests on the assumption that the tendencies to use or not use the two brain systems in optional ways are related only weakly, if at all. If people who often utilize top-brain processing do not often utilize bottom-brain processing, or vice versa, then we would have only Stimulator and Perceiver modes. We would not have Mover and Adaptor modes—but those, too, are posited by the new theory.

The findings about different imagery preferences for people in different professions can resolve this apparent contradiction. Here's how: What if the participants in the experimental studies already had different degrees of interest in humanities versus science, and it is this distinction that is responsible for the different imagery profiles? That is, perhaps the college students who served as participants had already specialized in disciplines that relied on one or the other sort of imagery, and engaging in the relevant sorts of thinking had led people to become good at spatial imagery and poor at object imagery, or vice versa.

So the question becomes: How are the two sorts of abilities related in a large population of randomly selected people?

A Harvard-based team consisting of Anita Woolley, J. Richard Hackman, Thomas Jerde, Christopher Chabris, Sean Bennett, and Stephen sought to find out.[12] These researchers had more than two thousand people take the Object and Spatial Imagery Questionnaire (with the special permission of the developers) over the Web; many of the test-takers were not students and not professionals. The figure on page 55 illustrates the results. Each dot represents the results from one person, with the scores on the two scales indicated by the height (for the object score) and position to the right (for the spatial score). The most striking feature of this graph is that there is only a very weak relationship between the two sets of scores.

The correlation measure, expressed as *r*, indicates how strongly two sets of scores are related. At one extreme, if knowing about one set of scores perfectly predicts another set of scores, the correlation between the two sets of scores would be $r = 1.0$ if the scores went up in tandem (as height and weight tend to do) or $r = -1.0$ if one score went up when the other went down (as days of starvation and weight tend to do); a correlation of $r = 0$ would indicate that knowing about one set of scores predicts absolutely nothing about the other set of scores. In this context, then, the observed correlation of $r = -.05$ is strikingly low. One way to understand the meaning of these correlation values is to square them; the result indicates how much of the variation in one set of scores is accounted for by variations in the other set of scores. For example, $r = .50$ squared is .25, which means that a quarter, 25 percent, of the variation in one set of scores can be predicted by observing variation in the other.

Knowing this, consider what happens when we square the *r* value of –.05: We get .0025, which means that a quarter of a percent of the variation in the spatial scores can be predicted by the variation in the object scores (and vice versa). For most purposes, then, the two sets of scores can be treated as not being related.

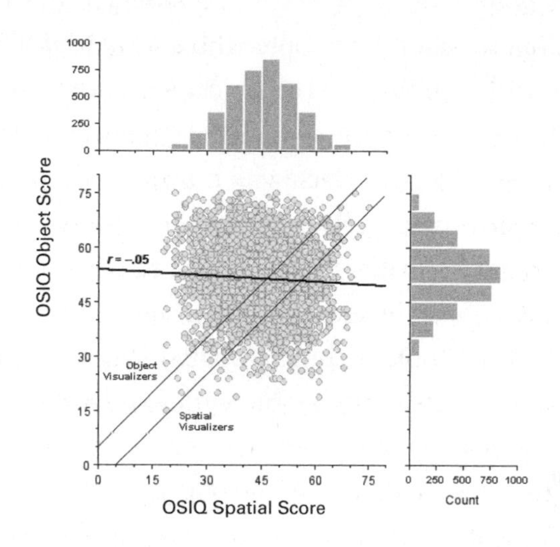

The relationship between scores on the Object Imagery versus the Spatial Imagery scales of the Object and Spatial Imagery Questionnaire (OSIQ). Each dot represents the scores on the two scales from a single person; the bar graphs illustrate how scores on the scales were distributed. Object visualizers had higher scores on the object scale than the spatial scale, and vice versa for spatial visualizers. The horizontal line shows how scores on the two scales were generally related. *With kind permission from Christopher F. Chabris.*

Boiling all this down, when a large sample of randomly selected people was examined, for practical purposes the two abilities were found to be independent. Yes, there is a small negative relation—indicating that people who are good at object imagery tend to be less good at spatial imagery, and vice versa—but when ordinary people are selected, a person's facility with one kind of imagery says almost nothing about his or her facility with the other kind.

The takeaway: As with the modes, people can be good at top-brain, spatial imagery and either good or bad at bottom-brain, object imagery, and vice versa.

However, a critical reader might wonder whether these findings just reflect random responses. Or perhaps the participants just guessed, knowing that they were anonymous on the Web. To assess this possibility, some of the people who had scored high on the spatial imagery scale (but not on the object scale) or high on the object imagery scale (but not on the spatial scale) were tested further. These people took two objective tests, which measured not preference but rather actual performance. Specifically, they took the mental rotation task (which requires spatial imagery) and the degraded pictures task (which requires object imagery, as described earlier).

As predicted, people with high spatial imagery scores did better on mental rotation than did people with low spatial imagery scores, and people with high object imagery scores did better on the degraded pictures test than did people with low object imagery scores. Moreover, spatial imagery scores did not predict performance on the degraded pictures test, and object imagery scores did not predict performance on the mental rotation test.

In short, the objective tests validated the questionnaire scores that were obtained over the Web. Clearly, the scores on the questionnaire *did* reflect underlying cognitive abilities; the scores were not just random.

We have made much of the top-brain/bottom-brain distinction. But what about that earlier, now-classic distinction, between the left and right sides of the brain? Is our distinction really better? In the following chapter we take up these questions.

Chapter 5

Sweeping Claims

We have introduced a new way to organize the brain into two parts—but what was wrong with dividing it into left and right halves? If there is something wrong with that earlier distinction, have we avoided the problems that beset it—or did we fall into the same trap? In this chapter, we review the history and status of the common left-brain, right-brain narrative.

This story apparently began on a day in February 1962, when a forty-eight-year-old man with intractable epilepsy was brought into an operating room at White Memorial Medical Center in Los Angeles. This was the moment that Roger Sperry, a world-renowned neuroscientist, had long awaited.

For years, Sperry and his colleagues at the California Institute of Technology had surgically separated the left and right sides (also called hemispheres because each is, roughly, half a sphere) of the brains of cats and monkeys, and then tested the animals in experiments that measured cognitive function. He and his team had developed a novel way to plumb the mysteries of the living mammalian brain—and their conclusions from this so-called split-brain research had taken the neuroscientific world by storm.

"They perceive, learn and remember much as normal animals do," Sperry wrote, in a paper that attracted significant attention inside academia but went virtually unnoticed in the larger world. "However, if one studies such a 'split-brain' monkey more carefully . . . one finds that each of the divided hemispheres now has its independent mental sphere or cognitive system. . . . In these respects, it is as if the animals had two separate brains."[1]

On that winter day in 1962, Sperry was setting the stage for his first test on a human, William Jenkins.

Jenkins was a military veteran who had suffered grand mal seizures ("brain spasms" that produce massive convulsions)—sometimes as many as ten a day—since surviving a bomb explosion near the end of World War II. He had learned of a radical operation, a version of which had been performed by other doctors two decades before at a Rochester, New York, hospital, that had relieved the symptoms of extreme epilepsy. He was eager for the California surgeons to try it on him. Unlike patients who underwent earlier versions of this operation, Jenkins made a deal with his doctors: Whether or not the surgery reduced his suffering, he agreed to work with Sperry, who would administer postoperative behavioral tests similar in principle to those given to the scientist's experimental animals. Assuming that Jenkins's higher cognitive functioning survived the surgery, his ability to respond on command and communicate with speech might give Sperry a quantum boost in his research.

"Even if it doesn't help my seizures," Jenkins said before meeting the scalpel, "if you learn something, it will be more worthwhile than anything I've been able to do for years."

The surgeons shaved Jenkins's head, sterilized and peeled back his scalp, opened two holes into his skull, and began the meticulous work of cutting the corpus callosum, the largest structure that connects the left and right hemispheres of the brains of humans; this structure consists of some 250 million nerve fibers, an impressive piece of brain anatomy. The operation went according to plan and Jenkins recovered without incident; his convulsions were indeed gone, and, like Sperry's monkeys and cats, on casual observation he seemed cognitively normal.

Six weeks after surgery, Sperry began to study his first split-brain human. Jenkins was gratefully cooperative during weekly sessions that continued for months.

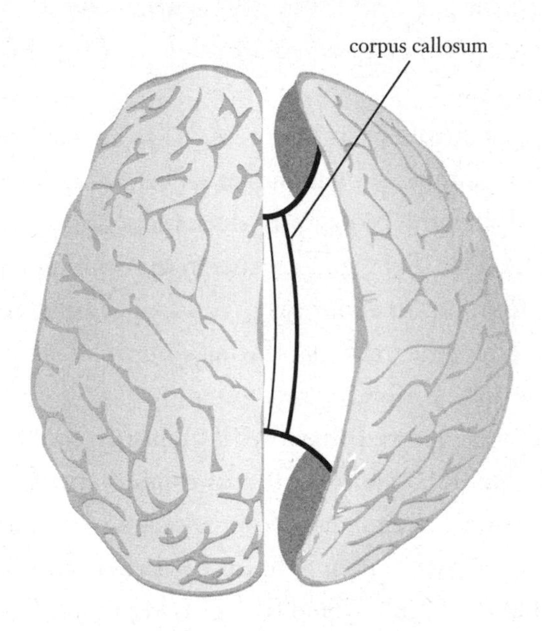

A view of a brain, seen from the top with a cutaway view that exposes the corpus callosum. The corpus callosum is the largest connection between the two cerebral hemispheres (left/right halves of the brain).

Sperry and his colleagues devised ingenious tests by which they could assess the cognitive functioning of each half of Jenkins's brain, together and in isolation. These tests relied on the established facts that the left hemisphere controls movement of the right side of the body (and vice versa), and the left side of each eye sends information to the left hemisphere and the right side of each eye sends information to the right hemisphere.[2] The results confirmed what Sperry had theorized from his studies of cats and monkeys: Each side of the human brain has distinct cognitive capabilities. Encouraged, Sperry and his colleagues pushed on, and by 1974, they had tested fifteen more people who had undergone surgical separation of the right and left hemispheres. The findings confirmed their pioneering conclusion: The two sides of the brain indeed do play measurably different roles in cognitive functioning.

By now, the larger world was paying attention. In 1981, Sperry was awarded the Nobel Prize in Physiology or Medicine for his split-brain research.

"The left hemisphere is the one with speech, as had been known, and it is dominant in all activities involving language, arithmetic, and analysis," the Nobel jury declared when awarding Sperry the prize. "The right hemisphere, although mute and capable only of simple addition (up to about twenty), is superior to the left hemisphere in, among other things, spatial comprehension—in understanding maps, for example, or recognizing faces."

Simple logic suggested that this dichotomy could explain why some people tend to be analytical and others tend to be intuitive. Some people soon extrapolated from Sperry's work and gave birth to a new theory of psychology built on the belief that the left side of the brain is the source of rational thought, logic, and linear thinking, whereas the right is the source of emotion, creativity, and imagination—and that in any person, one side or the other is dominant, making that person either more "left-brained" or "right-brained."

Inner Space

Sperry's groundbreaking research reached a wide audience at a time when the public had become fascinated with the brain in general. This was the heyday of print journalism, so when the multimillion-circulation *Life* Magazine began publishing a dramatically illustrated five-part series about the brain on October 1, 1971, ordinary people took notice.

"Miracles and mysteries on a tantalizing scientific frontier," the *Life* cover headline read, above a subhead that proclaimed: "With extraordinary photographs." And indeed, these photos were extraordinary: dozens of color images, many taken through an electron microscope, of vessels, tissue, neurons, and other parts. A remarkable world, existing inside each of us, was revealed. Hollywood had already brought moviegoers there, in the 1966 Academy Award–winning *Fantastic Voyage*, which spawned a succession of books, comics, and a TV series.

"It is the most highly organized bit of matter in the universe, this three-pound, electrochemical double handful of cells that thrives on change, allows us to move, see and think, to create, to love and be conscious of our actions," the *Life* editors wrote.[3]

> Since man first became aware of its existence, he has struggled to comprehend its miracles and miseries, punching crude holes in the bones that protect it and arbitrarily assigning moral and intellectual values to the lumps and bumps on its outer surface. His goals then and now have been the same: To gain a true understanding of how the brain works and use this knowledge not only to treat disease but to improve the very quality of life. Today, thousands of scientists from dozens of disciplines are pursuing these goals in the ultimate assault on man's last great scientific frontier on earth.

If Sputnik had inaugurated exploration of outer space, these researchers had embarked on a quest into inner space, as some had started to describe the frontier of the brain. But could they ever really succeed? With purplish prose, the *Life* editors wrought high drama from the field of neuroscience.

"There are difficulties—and dangers," the *Life* editors wrote.

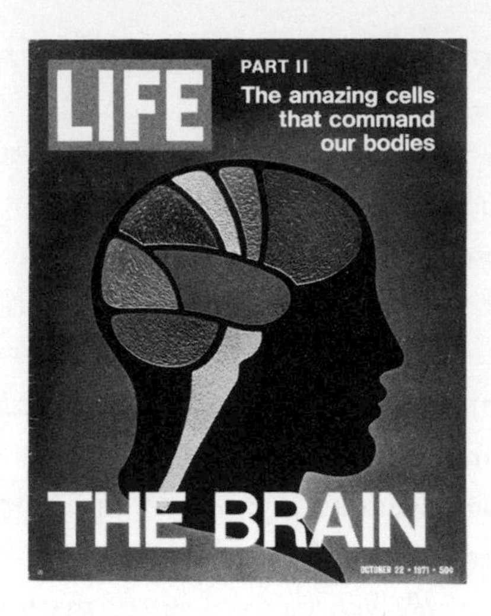

Life magazine's five-part series on the brain in the autumn of 1971 helped build public interest in neuroscience.

> After a decade of intense study, many basic questions remain unanswered. In fact, it is possible that the brain may be governed by principles too complex for it to grasp. And even if man does learn to dismantle the loom that spins out his existence, he will find himself with knowledge that could be misused.

The left brain/right brain story offered reassurance that we could be master of our own brains. This was fertile ground in which a new theory of psychology could take root.

Even before Sperry's Nobel Prize, the left brain/right brain story had started to spread through popular culture. It gained momentum two years after the *Life* magazine series, when the *New York Times Sunday Magazine* published an article, "We Are Left-Brained or Right-Brained."

"Two very different persons inhabit our heads," the article began, "residing in the left and right hemispheres of our brains, the twin shells that cover the central brain stem. One of them is verbal, analytic, dominant. The other is artistic." The article included a photograph of Sperry in his Caltech laboratory and a drawing of a brain divided into its two hemispheres.

The next year, the distinguished psychologist Robert E. Ornstein wrote the first of his many pieces exploring the concept, and his observations were featured in the July 8, 1974, edition of *Time* magazine. In 1976, the *Harvard Business Review* published "Planning on the Left Side and Managing on the Right," which instructed executives on how better to utilize both halves of their vertically divided brains. The widely circulating *Psychology Today* in 1977 splashed the left brain/right brain story across its pages. Countless other articles were published in specialized and general-interest outlets. A flurry of books began to appear.

Little wonder: The left brain/right brain story appealed both to those with an interest in psychology and to those who wanted only to improve themselves. The claims seemed to be based on solid science. They promised practical advice not only for personal matters but also for relationships and work.

This was not science for science's sake but a path to a better life that anyone could follow.

The Fine Print

Although he did indeed document profound differences between the functioning of the left and right cerebral hemispheres, Sperry did not intend his research to become the basis for a new branch of psychology.[4] In a 1984 essay published in the journal *Neuropsychologia*,[5] Sperry warned that "experimentally observed polarity in right-left cognitive style is an idea in general with which it is very easy to run wild . . . it is important to remember that the two hemispheres in the normal intact brain tend regularly to function closely together as a unit."

And, of course, there was the issue that the patients on whom all of the conclusions were based were just that—patients. They had abnormal brains (otherwise they would never have been operated on). So, there was a lingering question of just how strongly the results from these patients could be assumed to apply to normal people, whose brains are not split into two.

But that was the fine print.

Sperry's Nobel Prize, announced in October 1981, fueled headline-grabbing publicity. In October alone, the *New York Times* published ten stories about the scientist and his split-brain work. "On this foundation," the paper reported, "neuroscientists have postulated further hemispheric divisions of labor; that the left side, for example, may deal with logical, mathematical and analytic thought, while the right side may be the seat of artistic and musical ability."

Such notions about the left brain and the right brain built on each other and soon became a pop-culture avalanche; outside the research centers, the more nuanced and circumspect findings of neuropsychology and neuroanatomy were mostly lost as the popular version of the left/right theory became widely accepted.

Today, you can watch a YouTube video, as millions already have, of a "Spinning Dancer,"[6] and you will be asked to answer this question: "Do you see the dancer turning clockwise or anti-clockwise? If clockwise, then you use more of the right side of your brain and vice-versa." The Spinning Dancer even has its own Wikipedia page (in reality, the "test" is nothing but an optical illusion, and your designation depends on what parts of the figure you pay attention to; it was created by Japanese web designer Nobuyuki Kayahara).

Another self-test promises to reveal whether "your brain is right for a creative career."[7] Still another, which you can take at many sites, purportedly identifies "hemispheric dominance." Teens can take their own test, the results of which are alleged to be able to "improve your study habits."[8] If the author is to be believed, anyone can benefit from yet another test that assigns precise percentages of "brain usage"[9] to each side.

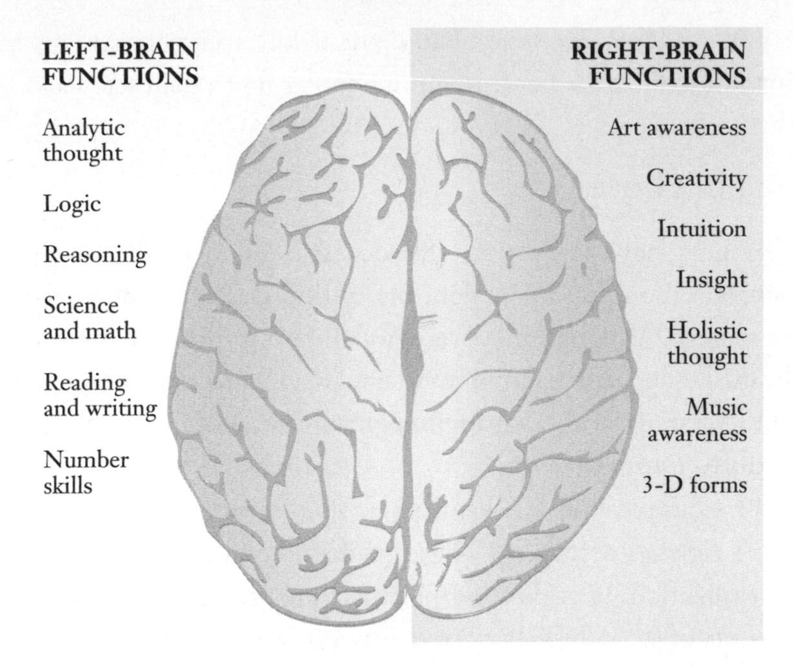

The major purported functions of the left brain versus the right brain, according to popular lore.

No one, it seems, is too young to benefit. Toys and DVDs can supposedly "develop" your toddler's left or right side (Stephen Hawking or Georgia O'Keeffe; you choose),[10] whereas an older kid might benefit from an ancient calculating device: "How can we motivate BOTH parts of the brain at a time? Learning abacus can accomplish this goal,"[11] one merchant asserts. And another Internet site claims that "whole brain integration means using the Left and the Right side of the brain together, which improves the use of your brain by a factor of 5–10%."[12]

Those inclined toward effortless improvement can even indulge in "essence therapy," as it is called: "Left/Right Brain Essence helps restore left/right brain balance," one ad promises. "Supports physical coordination, meditation, creativity, and mental and emotional balance."[13]

The problem is, these claims are wildly exaggerated—to the point where most of them are more popular myth than science.

Hand in Glove

Researchers have known for decades that none of the sweeping asssertions about left brain/right brain differences are supported by solid science. Although they were not shouting from the mountaintops, these scientists had unimpeachable evidence that the popular-culture versions of the left brain/right brain theory do not capture how the brain really works.

For example, the left hemisphere is often described as verbal and the right as perceptual—but this distinction doesn't hold up as a generalization. In reality, both hemispheres typically contribute to both sorts of activities—but do so, often subtly, in different ways.

Consider language: Typically, the left hemisphere produces correct word order—to say, for instance, "I have two left feet" instead of "I two left feet have." (Yoda's fractured English may indicate that his alien brain didn't include a human-standard left hemisphere.) But the right hemisphere also is crucial in language: It extracts the implied meaning—that the speaker doesn't literally have two left feet but has trouble with physical coordination, much as a person would if she were cursed with actually having two feet shaped like the left one (each with the big toe on the right and the smallest toe on the far left).

And although it is true that the left hemisphere controls speech and plays a major role in grammar and comprehension, the right hemisphere plays a key role not only in our comprehending implied meaning but also in our understanding and producing verbal metaphors and humor, and it is largely responsible for helping us to decipher the meaning of changes in speaking tone, such as the rising tone at the end of a spoken question. And *both* hemispheres play critical roles in extracting meaning in general. Indeed, neuroimaging studies have conclusively shown that many aspects of language processing are distributed over both hemispheres.[14]

Similarly, consider perception: For example, if you look at a house, the left hemisphere will allow you to pick up on the shapes of the doors, windows, and other parts, while the right will allow you to take in the overall contours of the building. At the same time, the left hemisphere will specify the relative locations of the parts in terms of categories, such as "the window is *left of* the front door," while the right hemisphere will specify locations in terms of specific distances, such as by indicating the precise distance the window is from the door. Again, brain imaging studies have conclusively shown that many aspects of perceptual processing are distributed over both hemispheres.

The larger issue is not just that people are being classified as "right-brained" or "left-brained" by so-called experts. It's that the hemispheres are being classified in terms of simple overreaching dichotomies—such as the left's being verbal, analytic, and logical, and the right's being perceptual, intuitive, and emotional. It just doesn't work that way.

Here are the two fundamental problems:

First, it is true that small areas of the brain are specialized in different ways in the two cerebral hemispheres, but these specializations are very specific. For example, a region near the front of the left hemisphere is adept at controlling the movements of the tongue, lips, and vocal cords during speech—but the corresponding part of the right hemisphere plays a crucial role in controlling such movements during singing. Similarly, a region of the left hemisphere under the temples classifies details of visually perceived objects, whereas the corresponding region of the right hemisphere classifies the overall shape of visually perceived objects. In addition, another region under the temples of the left hemisphere organizes speech sounds into the units of a familiar language, whereas the corresponding region in the right hemisphere organizes environmental sounds (such as the sound of rushing water or animal calls). And so on and so forth.

Although small brain areas sometimes do function differently in the two hemispheres, there may not be anything in common that characterizes *how* they function differently. For instance, what does the difference between controlling speech versus controlling singing have to do with the difference between classifying parts versus classifying overall shapes? So, when you start to group such small areas together into a larger area, any common thread soon breaks—and a simple dichotomy cannot characterize the larger area.

The sorts of documented differences between left-brain and right-brain functioning are hardly the stuff of popular generalizations, but they are fundamentally important to a genuine understanding of brain functioning. The fine print matters.

The second fundamental problem is that each of the specialized brain areas does not work alone but rather works as part of a system that includes many other brain areas—including areas on the opposite side of the brain.

To understand language fully, for example, you need to understand the syntax (the structure of sentences, which is better accomplished by the left hemisphere), the meaning of changes in tone (which is better accomplished by the right hemisphere), and how meaning is deciphered (which is accomplished by both hemispheres working together). In other words, the two hemispheres are part of a single system. Let's return to our example of a bicycle: It has handlebars, a seat, pedals, gears, a chain, and wheels. All of the parts are designed to work together to accomplish a specific goal (helping a person get from place to place quickly and easily). No one part alone would accomplish much; the power of the machine lies in how the parts all work together. The same is true of the brain.

So the hemispheres do differ, but at a more specific and detailed level than is claimed in the popular press and on the Internet. One half-brain is not "logical" and the other "intuitive," nor is one more "analytical" and the other more "creative." Both halves play important roles in logical and intuitive thinking, in analytical and creative thinking, and so forth. All of the popular distinctions involve complex functions, which are accomplished by multiple processes—some of which may operate better in the left hemisphere and some of which may operate better in the right hemisphere—but the overall functions cannot be said to be entirely the province of one or the other hemisphere.

And far from having separate lives, the two halves work together, as Sperry himself noted. They are not isolated systems that compete or engage in some kind of cerebral tug-of-war; one is not an undisciplined child, the other a spoilsport that throws schoolyard tantrums. Rather, as we have stressed, the brain is a single, marvelously complicated, and deeply integrated system. Like those of a well-maintained bicycle, the parts of the brain do have different functions—but, like the parts of a bike, they are designed to work together.

"Split-Mindedness"

If the pop-culture left-brain/right-brain story is off base, why has it persisted for nearly a half century? Why does a Google search bring millions of hits? Why have such respected figures as Oprah Winfrey[15] and best-selling author Daniel Pink[16] embraced it?

The answer is not only the intense media attention that Sperry's work and his Nobel Prize attracted or the power of the Internet today. The answer may lie in our instinctive search for understanding, a timeless narrative of the human experience. As a species, we seem to be hardwired to try to make sense of what we encounter, even something as complex as the brain—and so we create narratives, simplifying them when necessary. This is not inherently a bad thing, provided that the narratives are simplified in the right way—characterizing core ideas and not introducing misconceptions.

It is a testament to the power of the left/right story, and to our tendency to embrace simplified narratives, that this story has persisted despite periodic warnings from scientists and academicians.

As early as 1971, Brenda Milner, a distinguished scientist at the Montreal Neurological Institute, cautioned against overly expansive interpretations of split-brain research in an article published in the *British Medical Bulletin.*[17] In the conclusion to her study, Milner wrote that "although this paper has been written to emphasize the hemispheric differences, one must beware of pushing the contrast too far." Five years later, Stevan Harnad, founder (in 1978) of the respected journal *Behavioral and Brain Sciences*, wrote that a black-and-white lateralization dichotomy had "about as much relation to the known facts about hemisphere functioning as astrology does to astronomy."[18]

And in 1987, Pace University psychology professor Terence Hines maintained, in *Academy of Management Review*, that attempts to utilize the left brain/right brain story to improve business performance were like "the pursuit of wild geese," a pursuit "based on incorrect views of the nature of hemispheric differences, views that can be best termed *myths*"—his emphasis. The scientific literature holds many more such cautions.

But these contrarian voices were largely unheard—or unheeded—outside academia. Even for some inside the scientific world, the left/right theory proved tantalizing, as psychologist Ornstein remembered in his 1997 book *The Right Mind: Making Sense of the Hemispheres.* Recalling the experiments of Sperry and others in the 1960s and 1970s, he wrote:

> How these demonstrations of split mindedness got all our attention! It caused many people to overestimate the split of the mind and underestimate its unity, even though different sectors of the mind handle the world differently. In real life, one disconnected

> hemisphere isn't operating alone or running the whole show. The two are intimately and thoroughly connected, and not only by the corpus callosum, but by all the lower brain structures. And more.

We have argued that a person's habitual way of thinking *does* arise from the workings of two portions of the brain, the top and the bottom. And we have argued that simple dichotomies cannot adequately explain what these two portions do: They must be viewed as systems—and systems that work together. In many ways, the interactions between the top-brain and bottom-brain systems are more interesting and informative than the processing that occurs in each system alone. We turn to such interactions in the following chapter.

Chapter 6

Interacting Systems

Raised on a farm in New Hampshire, Phineas Gage demonstrated unusual resourcefulness at a young age. He received little formal education, but sometime in his teens or early twenties he must have concluded that he could better himself by forsaking his family's hardscrabble existence. Opportunity beckoned: This was the early Industrial Age, an era when railroads were booming, with new companies forming and new lines being built around the country. Gage learned the construction trade and aspired to management. The evidence suggests that he was a good planner who learned from his experiences and incorporated lessons learned in deciding every next step—a recipe for success. By 1848, he had advanced to foreman.

On September 13, 1848, Gage was in charge of a crew that was building a line for the recently chartered Rutland and Burlington Railroad that passed near Cavendish, Vermont. Late that afternoon, he was using a long, tapered metal tool called a tamping iron to pack explosives into a hole cut in rock. Momentarily distracted, he unintentionally let the iron hit the side of the hole. A spark ignited an explosion and the tamping iron shot upward from the hole like a massive bullet, passing through his cheek and head, destroying his left eye and taking out a significant part of the top and bottom front of the left side of his brain. Three feet, seven inches long, and an inch and a quarter in diameter at its thickest point, the tapered iron landed more than sixty feet behind him.

Gage was knocked down and may have briefly lost consciousness, but, amazingly, within a few minutes he was speaking coherently and able to walk. He was taken by oxcart to a nearby hotel, where local doctor Edward H. Williams was called to treat him.

"Doctor, here is business enough for you," Gage said when Williams arrived.

Not long after, Dr. John M. Harlow took charge of the case. Lying in blood on a bed, Gage pointed to the hole in his left cheek and told Harlow: "The iron entered there and passed through my head." It was a disturbingly precise description.

Harlow cleaned and dressed Gage's wounds—and was able, while searching for bone fragments, to touch his right index finger, which he had inserted through the hole in the top of Gage's head, to the index finger of his left hand, which he had inserted through Gage's fractured cheek. Harlow later wrote of what he observed: "The brain protruding from the opening and hanging in shreds upon the hair. . . . The pulsations of the brain were distinctly seen and felt."

Gage's convalescence was marked by infection and periods of coma; at one point, he was measured for a coffin. In spite of the setbacks, he survived and his speech, memory, and motor control were relatively intact. In late November, Gage returned to his family in New Hampshire. His case had attracted the attention of the press. Under the headline "An Astonishing Fact," one Boston paper published a letter from one of the many curious people who had seen Gage as he recovered in Cavendish. "We live in an eventful era," the writer stated, "but if a man can have thirteen pounds of iron in the shape of a pointed bar thrown entirely through his head, carrying with it a quantity of the brain, and yet live and have his senses, we may well exclaim, What next?"

But Gage was not the same man—and therein lay his appeal to the scientific community. Harlow chronicled a profound change in Gage's personality.

"Has no pain in head, but says it has a queer feeling which he is not able to describe," the doctor wrote after seeing Gage in April 1849, seven months post-accident.

> Applied for his [previous] position as foreman, but is undecided whether to work or travel. His contractors, who regarded him as the most efficient and capable foreman in their employ previous to his injury, considered the change in his mind so marked that they could not give him his place again. The equilibrium or balance, so to speak, between his intellectual faculties and animal propensities, seems to have been destroyed. He is fitful, irreverent, indulging at times in the grossest profanity (which was not previously his custom), manifesting but little deference for his fellows, impatient of restraint or advice when it conflicts with his desires, at times perniciously obstinate, yet capricious and vacillating, devising many plans of future operation which are no sooner arranged than they are abandoned in turn for others appearing more feasible. A child in his intellectual capacity and manifestations, he has the animal passions of a strong man.

What resulted was a sad and nomadic existence. With his brain crippled but still functioning and his left eye permanently closed, Gage was almost literally doomed to wander the earth. He first traveled around New England, exhibiting himself and his tamping iron, apparently for money. He was examined by Harvard surgeon Henry J. Bigelow, who wrote in an 1850 issue of the *American Journal of the Medical Sciences* that Gage had perhaps survived "the most remarkable history of injury to the brain which has been recorded."

He continued to drift, working for a while as an exhibit in P. T. Barnum's circus and spending several years as a livery worker in Chile, where a gold rush had attracted foreigners.

Alone save for his mother, he died in San Francisco after a series of seizures in 1860. His brain was not kept for dissection, but his skull and tamping iron wound up in Harvard's Countway Library of Medicine's Warren Anatomical Museum. His skull, with its distinctive hole, is today one of the images on the museum's home page, a macabre symbol of the brain's enduring power to fascinate and perplex.

Young Phineas Gage survived a gruesome accident that punished him cruelly but that led to rare insights into the living brain. *From the collection of Jack and Beverly Wilgus.*

Researchers usually interpret the sad tale of Phineas Gage as evidence that different aspects of personality rely on different parts of the brain. This is true enough, but the implications of his transformation are more profound than that: The tamping iron disrupted portions of the top and bottom parts of his brain, impairing how the top-brain system worked with the bottom-brain system. Gage had particular difficulty in integrating his emotional reactions, goals, and motives into his plans—and in knowing when to stick with a plan rather than allowing events to derail him.[1]

The damage did not simply disrupt certain capacities; it also changed the ways in which Gage's intact capacities subsequently were used. And his post-accident behavior did not simply reflect the fact that the top and bottom brain were damaged—it also reflected the abnormal ways that his top and bottom brain systems now *interacted.* By analogy, Richard Gregory in 1961[2] pointed out that when a resistor is removed from an old-fashioned radio, the radio may squawk. Why does this happen? It's not because the resistor was a "squawk suppressor," and removing it eliminated that function. Rather, the squawk occurs because the intact parts of the radio interacted differently after the damage. Similarly, after the accident, Gage's top-brain and bottom-brain systems changed how they worked together—and the changed interactions produced much of his altered behavior.

The disruptions of the normal interactions between the top- and bottom-brain systems devastatingly changed one major aspect of Gage's functioning. Wrote Harlow: "Previous to his injury, although untrained in the schools, he possessed a well-balanced mind, and was looked upon by those who knew him as a shrewd, smart businessman, very energetic and persistent in executing all his plans of operation. In this regard, his mind was radically changed, so decidedly that his friends and acquaintances said he was 'no longer Gage.'"

Whereas Gage previously had been strategic and thoughtful, he now was impulsive and unstable. His bottom-brain system interrupted his top-brain system inappropriately, impairing his ability to stick to plans or revise them when he learned the consequences of previous efforts. When his reactions disrupted his plans, he was left awash in a sea of fluctuating emotions and was incapable of responding appropriately. The disruption in how the two brain systems interact had altered how Gage related to other people and how he behaved in the daily situations he encountered.

Systematic Interactions

We have emphasized the importance of thinking of the brain as a system, with inputs, outputs, and specialized processors that produce appropriate outputs for inputs. We've also stressed that the top and bottom portions each comprise specialized smaller systems that work together as parts of the greater whole. It is tempting to use the analogy of a computer—and, to a certain extent, the analogy fits. That is to say, brains and computers both process information: In both cases, input (such as from the eyes to the brain, or from a camera to a computer) is stored, transformed, and manipulated, and eventually produces an output (such as a spoken word or an image on a screen).

But brains and computers differ in *how* they process information. A computer has a separate "central processing unit," the hardware of the machine that carries out the instructions of programs; separate random-access memory (RAM, which can be increased by adding more memory chips), for quick access to data; and disk or solid-state storage, the machine's "secondary" or "storage" memory, where data and programs are stored (this kind of memory retains information even when the power is off). Brains, by contrast, don't have a clear distinction between a central processing unit and memory—and brain structures used to store information over long periods of time (analogous to a computer's disk or solid-state storage) may also be involved in storing information for brief periods (analogous to a computer's RAM).

Nonetheless, an insight gleaned from computers is the fact that information processing can be understood only within the context of a system with multiple coordinated components. To illustrate the intricacies of how the top- and bottom-brain systems interact, we can use an analogy of a commercial bakery with two floors.

Let's say that it's the week before Thanksgiving. The bakery needs to produce more pumpkin pies because consumer demand is rising, as it does every holiday season. On the top floor are the executives who plan how many pies and other baked goods to produce. Their plans need to take into account various sorts of information, such as the season, day of the week, and availability of specific ingredients (for instance, pumpkins). They then place orders for the ingredients. As the big day approaches, they monitor sales, advance orders, and other indicators, and adjust how many pumpkin pies should be produced. The executives formulate plans based on their expectations, execute them, and then revise their plans as new information arrives.

At the same time, on the bottom floor, many people check that the pumpkins, flour, sugar, and other ingredients arrive, sort them, ensure that the ingredients are fresh (and discard any spoiled ingredients), send them to the appropriate mixers and ovens, and so on. They organize what arrives from the outside world, sort it into categories, and interpret what should be done.

If the floors did not interact, no pumpkin pies (or bread or other baked goods) would be produced. The point is: They *do* interact. The plans formulated on the top floor are relayed to the workers below, so that they are ready to receive certain ingredients and monitor certain information; and the results of the baking efforts and the information being tracked below (including information about sales) are sent back to the top floor, so that the executives can discover how well their plans are going and adjust them accordingly.

What happens when sales of pumpkin pies are not as good as expected? This information, monitored below by salespeople, is relayed to the executives above. These top-floor employees then scale back how many pies to make the next day, accordingly ordering fewer pumpkins and other ingredients. The workers below would be told to expect less of each ingredient and would be prepared to process the altered amounts. They then let the top floor know the amounts that actually arrived. If the amount of a particular ingredient that they report is more or less than expected, the executives would contact the suppliers of the relevant ingredient (pumpkins, flour, sugar, etc.) and have words with them—ensuring that the bakery didn't have to pay for more than they requested if too much was shipped or they had sufficient ingredients to fill orders if the shipments were short.

Bringing this back to the brain, consider the following example: You want to go online, so the top-brain system first formulates a plan to turn on your computer and, after it's on, access your browser. After you turn on the computer, the top-brain system expects to see the sign-in screen; when it appears (and is registered by the bottom-brain system), the top-brain system sets up a plan for entering your password when you start typing. The top-brain system not only generates the commands to control your fingers but also produces expectations about what you should see as each letter appears. It receives input from the bottom brain about which letters appear, and the top-brain system notes if an unexpected letter appears and—if so—revises the plan and corrects the mistake.

This is not to minimize the contributions of the bottom-brain system. When you see a computer screen, the bottom-brain system organizes the pixels into patterns that correspond to words and pictures; it then compares these patterns with all the stored information about things you have seen before; if it finds a match, it applies to the present case the information that you previously associated with the identified object or pattern. Consider your reaction when you see the symbol ⃠: you know that it means "prohibited" because you've seen the symbol before and its meaning has been stored in your memory. When your bottom-brain system matches the input from your eyes to this stored pattern, you then can apply the information you previously associated with it to the present pattern. But more than this, the emotional valance associated with the object that produces the input helps you decide on priorities. If you encounter something that is inherently valued (for example, a hundred-dollar bill on the sidewalk) or aversive (dog droppings on the sidewalk), your ongoing behavior may be interrupted in order for you to pursue a new priority (bending down to pick the money up or moving to the side to avoid stepping on the dog waste).

Moreover, it's not just that the top system uses the output from the bottom system as part of how it receives feedback about the consequences of acting on a plan; the expectations produced by the top system can bias the bottom system so that it is likely to classify inputs in a specific way. That is, the top brain adjusts the bottom brain so that it can easily perceive what is expected. For example, to a farmer gathering his cows at dusk, even a passing shadow of an appropriate size may be classified by his brain as a cow. Why? The farmer's expectations, created by the top-brain system, biased the bottom-brain system to classify input as a cow—and biased it so strongly that relatively little input is sufficient to qualify as the expected object.

In addition, the bottom brain draws on the top brain in interpreting the world. The bottom brain automatically classifies objects and interprets scenes and unfolding circumstances by matching them to information previously stored in memory. But sometimes we encounter objects and events that, in combination, don't match anything familiar; we may be familiar with the individual objects but not the way they are combined.

Say you saw a guy jumping around on one foot, wearing socks over both hands, and singing "America the Beautiful." Your bottom brain would register each of these things and send the information to the top brain, which then would try to generate a narrative to make sense of the ensemble. Perhaps this is a fraternity initiation rite. The top brain might then generate a plan to confirm or reject this conjecture, perhaps by looking for young men nearby who could be fraternity brothers monitoring the situation. And in so doing, the top brain would prime the bottom brain, making it easier to see the predicted brothers. Without question, the two systems interact.

Depending on the situation, this sort of processing can happen over an extended period of time—or faster than the blink of an eye. Let's look at the cognitive functions involved when a jet pilot returns to an aircraft carrier: To the aviator, the flight deck appears to be about the size of a postage stamp as the jet approaches. The pilot gently shifts the plane's controls, slowing it down and initiating descent (all of this in response to top-brain plans). In so doing, he or she expects to see the deck change in specific ways—for example, it should loom larger as the plane approaches it. But if this change (as initially registered in the bottom-brain system) is not as expected (registered in the top-brain system), the pilot will revise the plan. Say a crucial control was set wrong or malfunctioned and the plane did not begin to slow. The bottom-brain system would register that the flight deck is looming larger much more quickly than expected, and would immediately relay this information to the top-brain system. In response, processing in the top-brain system would lead the pilot to check the controls and try to locate the cause of this malfunction. She or he might have to abort the landing, regaining altitude as flight controllers (their own brains engaged in top-bottom interplay) worked to help to resolve the situation.

Such split-second decisions a pilot makes in executing a landing are a testament to the two brain systems' extraordinary and interacting capabilities. Without the systems' working together, the pilot never could have gotten the aircraft into the air, never mind returned safely.

• • •

As we saw in chapter 3, numerous connections run between the bottom brain and top brain. For example, a large bundle of nerve fibers called the arcuate fasciculus runs from the temporal lobe to the parts of the frontal lobe involved in producing speech. And numerous connections run to and from areas of the temporal lobe that store memories and various parts of the top-brain system. Clearly, the two brain systems are part of a larger, integrated system. As we shall see in the next chapter, the interactions between the top-brain and bottom-brain systems produce the four modes of thinking that will be our focus in the remainder of this book.

Chapter 7

Four Cognitive Modes

The story of Phineas Gage is now regularly included in textbooks about the brain, usually to illustrate functions of the parts of the frontal lobe and their specific contributions to personality. Rarely do textbooks note that the consequences of this damage arose because the top-brain and bottom-brain systems no longer interacted appropriately. We have focused on this fact.

But more than that, we explain the effects of Phineas Gage's injury in a new way, through the lens of the Theory of Cognitive Modes: When the tamping iron ripped up part of his brain, it changed his dominant cognitive mode, changing how he approached the world and interacted with other people.

Differences in how well the top-brain and bottom-brain systems interact do not arise only after a brain injury. Although all of us use both systems of our brains (if we didn't, we couldn't function in daily life), within the normal bounds some people may rely on each system to greater or lesser degrees. People differ in how much they rely on the two systems, and hence how the two systems interact. And differences in the interactions between the top- and bottom-brain systems produce different cognitive modes.

Our theory does not posit that people differ in the *sizes* of their top or bottom brains. Moreover, the fact that someone often relies on a brain system does not imply that he or she uses that system effectively. For instance, he or she may often rely on the top-brain system, devising many plans—but the plans may not be very good. How effectively a person uses the top-brain system is probably related to intelligence, which is distinct from the modes. People may in fact differ in the sizes of the top brain and bottom brain, and they may in fact differ in how effectively they use one or the other system—but this is not what the Theory of Cognitive Modes is about.

Rather, the cognitive modes emerge from the *degree* to which a person relies on the top- and bottom-brain systems. As noted earlier, we all constantly rely on both the top and the bottom systems, but we may often engage in only the minimal amount of processing necessary to respond to a situation. At other times, we engage the systems in optional ways. By analogy, we may need to walk to get somewhere (the brain responds to the situation with minimal processing), but we never need to dance (dancing involves an optional use of the brain). It is the optional sort of utilization that we focus on here—the extent to which a person uses a brain system not when forced to do so by the situation, but rather because he or she has developed ways of relating to the world and others that depend on using one or both brain systems.

In what follows, when we talk of the top- or bottom-brain system's being utilized or relied upon, we mean in this second way—optionally, when others might not use the system in this way.

The degree to which the systems are utilized in this second sort of way forms a continuum—ranging from highly utilized to minimally utilized—but for practical purposes, we can divide the continuum into "high" and "low" categories. As we will discuss in the following chapter, these differences (like most other cognitive, emotional, and behavioral characteristics) probably arise from an interplay of inherited characteristics and personal experience.

With two brain systems and two possibilities for each of them, we can thus identify four different cognitive modes—four different ways of interacting with people and responding to situations that arise in the world:

	Highly Utilized Top	**Minimally Utilized Top**
Highly Utilized Bottom	Mover Mode	Perceiver Mode
Minimally Utilized Bottom	Stimulator Mode	Adaptor Mode

Mover Mode results when the top- and bottom-brain systems are both highly utilized.

According to our theory, when people think in this mode, they are inclined both to implement plans (using the top-brain system) and to register the consequences of doing so (using the bottom-brain system), subsequently adjusting plans on the basis of feedback. The evidence suggests that prior to his injury, Phineas Gage often relied on this mode when he was at work; he probably could not have risen so far so fast if he had not. But after his accident he could no longer operate in this mode.

People who habitually operate in Mover Mode tend to be well suited to being leaders. They might head a company, act as a principal of a school, or take charge of revising a church afterschool program. According to our theory, people who habitually operate in this mode should be most comfortable when in positions that allow them to plan, act, and see the consequences of their actions.

You may know someone who habitually operates in Mover Mode. Perhaps she is the head of a neighborhood association. This person consistently looks ahead and devises plans, which she puts into action. For example, she may be the one who comes up with a clever way to get businesses to donate services for the annual fund-raising auction. But she does not blindly charge ahead. If a plan to have this fund-raiser falters, for example, she would be the first one to think about what went wrong and how to do it better next time.

Perceiver Mode results when the bottom-brain system is highly utilized but the top-brain system is not.

People who highly utilize the bottom-brain system try to make sense in depth of what they perceive; they interpret what they experience, put it in context, and try to understand the implications. They may use the top-brain system to generate narratives that make sense of what the bottom brain registers, but they do not use the top brain to initiate complex or detailed plans; the top brain is largely used in the service of the bottom brain. Gage would have been better off post-accident if his top-brain system had been better able to sort the inputs from his bottom-brain system.

Among others, many librarians, naturalists, and pastors appear to rely habitually on Perceiver Mode. If the Theory of Cognitive Modes is correct, people who are relying on this mode often play a crucial role in a group; they can make sense of events and provide a bigger-picture perspective. In business, they are often crucial members of teams, providing perspective and wisdom but not always getting credit.

To use our neighborhood association example, someone operating in this mode may be quiet during the meeting—but she is listening intently, and clearly tracking what's going on. Until she has something well founded to say, she keeps her own counsel—but she isn't shy about speaking up, once she's sure she has something to say. And because she deeply understands what she hears, she's often worth listening to. For example, when she does speak about the fund-raiser plans, everyone listens; if she thinks she's spotted a flaw (for example, she thinks that the marketing message might alienate some families), she probably has good reason to think so.

Stimulator Mode results when the top-brain system is highly utilized but the bottom-brain system is not.

People interacting with the world in Stimulator Mode often create and execute plans (using the top-brain system) but fail to register consistently and accurately the consequences of acting on those plans (using the bottom-brain system). They may be creative and original, and may be able to think outside the box even when everybody around them has a fixed way of approaching an issue or situation. But, at the same time, these people may not always note when enough is enough—their actions can be disruptive and they may not adjust their behavior appropriately. Gage's problem was not so much in registering the consequences of his actions but in too freely allowing ongoing situations to interrupt his plans; the damage disrupted the usual interactions between his top-brain and bottom-brain systems.

In general, when people think in Stimulator Mode, they should be able to play a crucial role as a team member; however, to be most successful, they should not be the sole leader but would be better off working with others who can help them adjust their plans as events unfold.

You may know someone who habitually operates in this mode. She may be a member of a neighborhood association committee—the woman who throws out an idea a minute with happy abandon. You might be tempted to dismiss her out of hand, but some of the ideas are good—even though she herself makes little effort to sort through them. If she is put in charge of a project, she probably is as likely to fail as to succeed—not necessarily because the basic idea of the plan is bad, but rather because she doesn't stay on top of fine-tuning it as events unfold.

Adaptor Mode results when neither the top- nor the bottom-brain system is highly utilized.

People who are thinking in this mode are not caught up in initiating plans, nor are they fully focused on classifying and interpreting what they experience; instead, they are open to becoming absorbed by local events and the immediate requirements of the situation. If the Theory of Cognitive Modes is correct, they often are "action-oriented" and responsive. In addition, people who habitually operate in this mode often "go with the flow" and may tend to be seen as free-spirited and fun to be with. (Gage would have been better off if he had been able to adopt this mode and let others define his agenda.)

People who are thinking in this mode should be valuable team members because they can easily adapt to plans. In business, people who typically operate in Adaptor Mode would often form the backbone of the organization, carrying out the essential operations.

Someone who habitually thinks in Adaptor Mode will not dominate formulating strategy during the neighborhood association meeting and may not have much to contribute during this planning stage. But once plans are in place, she embraces the assignment and works hard to carry it out. If asked to go door-to-door to solicit services from businesses for the fund-raising auction, she typically will be happy to do this. If the plan doesn't work so well (few businesses respond positively), she won't make much of an effort to figure out how to fix the problem—she's already carried out her role.

Context and Comfort

We now must note some caveats regarding the four modes of thinking.

First, it's important to realize that although the predictions we just summarized flow naturally from the theory, we have not directly tested them. Nevertheless, the questionnaire that assesses dominant mode, presented in chapter 13, has been shown to be reliable (comparable scores are usually obtained when people take it a second time)—and its reliability suggests that people do in fact behave consistently in many situations and that each of us does have a dominant mode. And, as we summarize later, the test has easily interpretable statistical relations to other, well-validated tests.

Second, as we noted earlier, the habitual mode of thinking generally should *not* be related to intelligence. Intelligence is about how quickly and easily a person can solve difficult problems and understand complex material. People who rely predominantly on one or another of the modes can be very intelligent, not very intelligent, or anywhere in between. The modes are about how you approach or interact with other people and situations that arise in the world; they do not reflect your ability either to solve problems or to understand complex material.

Another caveat is that we do not expect people who habitually operate in any particular mode to be more or less prone to be emotional or to have better or worse mental health; for instance, people who habitually operate in any of the four modes should be equally inclined to become angry or depressed.

Furthermore, even though we may habitually rely on one mode, each of us nevertheless may sometimes adopt different modes in different contexts. For example, if you were ever head of a neighborhood association, in that context you may have adopted Mover Mode. But if you were thrown into an unusual situation (such as being asked to devise a new way to motivate your neighbors to recycle their trash), you may have slid into Stimulator Mode—throwing out ideas without a clear way of knowing how to react to the results. To deal with complications, you may have adopted Perceiver Mode and sat back to observe before deciding to do anything. Finally, in certain situations, such as those dominated by other people who were very effective in Mover Mode, you may have decided to be a good team player—and relied on Adaptor Mode.

As we've noted before, according to the Theory of Cognitive Modes you are not "trapped" by your dominant mode, even though it is the mode you are most comfortable using (because it is most consistent with your temperament and experience, as we discuss in the following chapter). Nevertheless, we will argue that although people sometimes operate in different modes in different circumstances, we each have a dominant habitual cognitive mode that guides much of how we approach life and behave.

• • •

Identifying which mode you typically operate in can lead to self-insight. However, we are not suggesting that you try to change yourself based on these insights; rather, it might be best to use them to improve how you relate to your work and life more generally.

Because the theory we present here should help you to identify another person's dominant cognitive mode by observing his or her behaviors, this theory can also help you better understand how and why others act as they do. This could be of practical benefit to you in a relationship, at home, in social circumstances, and on the job.

Let's say you work in the production department of a medium-size advertising firm. Your company has hired a new head of social media, Sara. Sara seems frequently to operate in Perceiver Mode: She is an acute observer and has a keen understanding of human nature, as her tweets and Facebook wall postings confirm. But she's having difficulty putting all this together into a coherent campaign for the cable TV account your company has just landed. You, on the other hand, have identified yourself as typically operating in Stimulator Mode: You're good at drawing up plans, though not as good at reacting and adjusting wisely as your plans are put into action. So far, the company president has not paired you and Sara on the cable TV account. But wouldn't that pairing make sense for each of you—and your firm? Maybe you and Sara should present yourselves to the president as a team. By combining forces and drawing on your respective strengths in Perceiver and Stimulator modes, you might be able to make the whole more than the sum of its parts.

Or let's say you're someone who typically thinks and behaves in Adaptor Mode. People like hanging around with you because you often seem to live in the moment. You're easygoing and fun. What hasn't been so much fun, however, is the relationship with your partner, who also typically relies on Adaptor Mode. You have just broken up, again, in part because the two of you have difficulty making plans—and, frankly, trouble figuring out just where you are headed. Your disagreements ended in more disagreements, not lessons learned, at least until now, when you have finally agreed to go your separate ways. When you are ready for another relationship, it might be wise to consider more carefully someone whose dominant cognitive mode is not the same as yours—perhaps someone who habitually relies on Mover or Perceiver mode.

Note that we are not recommending that you use our theory to guide your life—only that you use it as an impetus to consider situations from new perspectives. As these examples illustrate, the Theory of Cognitive Modes should help you reflect and consider your options. In later chapters, we describe possible strategies for coping if you find yourself having to function with someone whose dominant cognitive mode rubs you the wrong way, whether at work or in a personal relationship. We hope that such speculations will help you see new ways to think about yourself (such as the sorts of situations in which you might need help, and to whom to turn for particular kinds of help), the way you relate to work, and the nature of your relationships with other people. The simple act of thinking about such things from a new perspective could prove illuminating.

But what if you don't like your dominant mode of thinking and want to change? How difficult would that be? In order to answer this question, we need to consider why people develop the dominant cognitive mode they do.

Chapter 8

Origins of the Modes: Nature Versus Nurture

Although evidence suggests that each of us tends to rely on a particular cognitive mode, no one is frozen into one mode of thinking at all times. People sometimes shift their modes of thought, and that is why we deliberately use words such as "typical" and "habitual" throughout this book. According to the Theory of Cognitive Modes, your dominant mode (i.e., the one you usually operate in) is determined by how much you utilize the top-brain and bottom-brain systems (where "utilize" refers to the optional uses of the systems, as discussed previously)—and you can alter this, albeit with considerable effort.

Consider one of the great diplomats of our time: the late Richard Holbrooke, who served several presidents (and also succeeded on Wall Street) and is perhaps best remembered for brokering the 1995 Dayton peace accords, which ended the bloody war in Bosnia. At the time of his death in December 2010, Holbrooke was a special adviser to President Obama on Afghanistan and Pakistan. Holbrooke understood not only history and contemporary geopolitics but also the complicated personalities of the leaders who drive them. Clearly, he often operated in Perceiver Mode. But a successful negotiator on the world stage must also, of course, sometimes operate in Mover Mode—and Holbrooke certainly did, as Dayton and other achievements demonstrated. And yet, in his well-known angry rants, he sometimes appeared to be in Stimulator Mode. These outbursts sometimes worked to his advantage (and hence may have been strategic, reflecting Mover Mode thinking), but they also, in certain circumstances, alienated others and apparently were not part of a plan to accomplish a specific goal.

Or consider Stephen Colbert, the popular comedian who plays a right-wing talk-show host on television: He himself must often be in Mover Mode in order to project a character who is usually supposed to be in Stimulator Mode when he interviews a guest, throwing out bombshells and being intentionally provocative in order to "see what happens." But in order effectively to see what happens he must sometimes shift into Perceiver Mode. If he notices that the guest is getting upset in a way that isn't very entertaining, he might shift into Adaptor Mode and let the guest take control of the interaction for a while.

Stephen Colbert himself is obviously very comfortable in Mover Mode, but he can let himself—at least temporarily—function in the other modes. So it is with many of us as we move through our less public lives.

DNA Meets Experience

Nonetheless, if the modes of thinking are like other psychological characteristics that have been studied in detail, people fall into a single mode by default for specific reasons—and some of these factors are difficult to overcome. In particular, a person's temperament—which is one aspect of personality—probably affects why that person habitually operates in a given mode; some aspects of temperament affect how easily the person can stay focused on executing a complex, detailed plan or reflecting on the interpretation of an event.

Such aspects of temperament include how emotional a person is, his or her overall activity level, his or her attention span and ability to be persistent, and how reactive and patient she or he is. Consider Perceiver and Stimulator modes, for instance: Being patient would help one to operate in Perceiver Mode but would not help one to operate in Stimulator Mode—but the opposite would be true for being highly active, which would help one to operate in Stimulator Mode but not in Perceiver Mode.

Why do we have the particular temperament we do? Most aspects of personality are determined at least in part by genes. Thus, it is plausible that genes influence which mode we prefer. And you cannot easily change the aspects of your temperament that were programmed by your genes.[1] The genetic effects of temperament are evident even in infants. If you are a parent, you will have noticed that even as babies, children are different: One might be alert and lively, another a calm and relaxed mini-Buddha, and a third seemingly high-strung and jittery.

Studies of twins have produced good evidence that temperament is given to us—at least in part—courtesy of our genes. Because identical twins have almost the identical set of genes and fraternal twins share only half of their genes, researchers can infer the role of genes by comparing the two groups. Studies have consistently shown that many aspects of temperament are more similar in identical twins than in fraternal twins. Such studies have implicated genes in many aspects of temperament, including activity level, level of attention and persistence, emotionality, and shyness. The estimates of "heritability," which corresponds to the percentage of the variability in a characteristic that is due to genes, range from about .20 to .60 (on a scale from 0 to 1.0). Without question, temperament is determined in part by genes.

Consistent with such empirical findings, researchers report that many aspects of temperament are relatively stable as children grow up. This fact has been well illustrated in the work of a pioneer in this field, the Harvard psychologist Jerome Kagan.[2] He has focused in large part on a facet of temperament he calls "reactivity," which is evident even in four-month-old infants. Kagan and his colleagues showed infants unfamiliar objects, such as a strange-looking metal robot, and had them take part in novel procedures, such as being fitted with a blood pressure cuff. About 20 percent of the babies he and his colleagues tested were highly reactive—they became agitated and aroused when they encountered unfamiliar objects or events. And roughly another 40 percent were not highly reactive, but instead were relaxed and didn't react strongly to unfamiliar events.

Years later, these same children were tested again to assess their brain functioning. And in fact ten- to twelve-year-old children who were high-reactive infants were found to have different brains from most other people.[3] Specifically, these children had more active amygdalae than the children who were low-reactive as infants. The amygdala is a deep-brain structure that contributes to our having strong emotions, such as fear. At least in part, its future functioning is decided at conception, when a person's DNA is assembled.

In short, temperament probably affects our dominant mode, and genes clearly influence our temperament. Our genes are not only part of the reason why we look as we look—they also are part of the reason why we behave as we behave.

You are also stuck with baggage that is not a result of your genes. Learning is a big factor that affects behavior—and we have good reason to think that learning affects which cognitive mode you find most comfortable.

In most cases, we cannot help learning and remembering information about events that we pay attention to—that's simply how the brain works. That is, we don't have to make a deliberate effort to remember most things; merely paying attention to objects or events as we encounter them, and thinking about them at the time, typically will lead us later to remember them. This characteristic of the brain causes many sorts of important experiences to affect us for years to come. For instance, just being at the wedding of a close friend, and paying attention to every detail, will likely lead you to remember many happy aspects of the event many years later. By the same token, being at the funeral of a loved one, and hanging on every word of the eulogies, will likely lead you to remember the unhappy event—whether you want to or not.

Moreover, it's not just genes and learning acting separately—it's the way they interact that uniquely defines who you are and what you can do. If your genes incline you toward a very active temperament, you will probably engage in sports and take trips that involve an element of adventure—perhaps going rock climbing on very challenging mountain faces or rafting down rapids. But if your genes incline you to have a more passive temperament, you will likely gravitate toward more laid-back pursuits—reading or gardening, for example. And, once you've had those experiences, what you learned will in turn shape you, giving you new ways of organizing the world and classifying future experiences (relying on the bottom brain). Moreover, they will expand your repertoire of viable plans, of ways to go about doing things (relying on the top brain).

In short, your genes may incline you toward different experiences, and the learning that occurs during those experiences in turn affects how much you utilize the top- and bottom-brain systems.

A telling finding comes from the study of IQ: The genetic contribution toward IQ actually goes up as people advance into old age, as University of Minnesota researcher Thomas J. Bouchard, Jr., reported in 2004.[4] Why would this be? One reason is that as you age, you have more control over your environment: You get to decide whether to spend time hunting deer or cuddling up with a good book, playing touch football or interacting with other chess players, and so on. Because the effects of your genes (such as those aspects of temperament that are genetically programmed) weigh in on your choice of environments, and environment contributes to IQ, your genetics can affect your IQ indirectly. The indirect effects may increase with age because you are able to choose your environment, and thus the total contribution of the genes to your intelligence increases with age.

We can credibly conclude that our genes almost certainly influence the degree to which we each tend to use our top- and bottom-brain systems in optional ways, not forced by the current situation. And we can conclude that learning, which affects virtually all aspects of cognition and behavior, also influences how much we utilize the two systems.

It follows that over time each of us tends to settle into using the top- and bottom-brain systems to greater or lesser degrees—which in turn produces a dominant cognitive mode. We may not be stuck in this dominant mode every minute of every day, but we will be most consistently comfortable using that one particular mode.

Changing Modes

So, what to do when your habitual mode of thinking turns out not to be as helpful as it could be for the job at hand? Can you change which cognitive mode you use? According to our theory, the answer is: In most cases, yes—but only to a degree, and only with significant effort.

In what follows we'll examine the possibility of changing your dominant cognitive mode, examining ways to alter the functioning of both the top- and the bottom-brain systems. We'll start with the bottom brain, which classifies and interprets input from the senses.

If experience has made you an expert in a certain area—you've learned a huge amount about it, in other words—you can become comfortable using Perceiver Mode in that area, regardless of your dominant mode of thinking. The problem is that the scientific literature indicates that, in general, people become experts only after about ten thousand hours of practice. Unlike cognitive modes that are determined in part by temperament—which may affect all your interactions with the world—cognitive modes that emerge from *knowledge* in an area are typically restricted to that one domain. If you become an expert in football plays, you may become very good at classifying and understanding them—but this skill will not help you much, if at all, with classifying and understanding baseball or soccer plays. In fact, in many cases becoming good at one particular task does not even lead one to become good at *similar* tasks.

Consider chicken sexing, an important practice in animal husbandry. Determining whether day-old chicks are male or female is crucial to egg producers: Not only are males incapable of producing eggs, but their presence actually disrupts egg laying. Poultry farmers place a premium on sorting males from females.

Classifying the sex of baby chicks is easier said than done. To the untrained eye, it is not at all clear what distinguishes males from females—they don't have easily identifiable external genitalia. This explains why becoming a professional chicken sexer is a valued, high-paying profession; these experts were in demand even during the Great Depression. Expert chicken sexers can determine the gender of about a thousand chicks per hour with 98 percent accuracy. In about half a second, looking under a magnifying glass, they make the call. Imagine a baseball hitter with even half this rate of success—he would surpass even Babe Ruth and Ted Williams, the greatest hitters who ever lived, who got a base hit only about a third of the time at bat.[5]

Researchers Irving Biederman and Margaret M. Shiffrar studied how chicken sexers perform their job.[6] They began by asking them to look at photographs of baby chicks' bottoms and circle the area that they focused on when determining sex. The area that was identified was flat or concave for females and convex for males. The researchers then used photos to train people with no experience in this domain. After training, the nonprofessionals could distinguish males from females accurately 84 percent of the time. (Actual training typically does not involve photos but rather hinges on experience with real chicks.)

This study illustrates two important points:

First, the task could be learned—after instruction, participants did far better than the 50 percent accuracy expected by chance alone. Even without much practice, the nonprofessionals were able to function in Perceiver Mode when classifying chicks.

Yet the training did not allow participants to generalize completely, even within this one domain: Those trained in the study were not even close to being perfectly accurate; they missed many calls. In fact, it turns out that to be an expert chicken sexer requires mastering various exceptions to the simple rule. Learning the rule doesn't allow a nonprofessional to generalize, any more than learning that a strike in baseball is a fruitless swing of the bat helps you also understand that a strike occurs when the ball passes over the plate in a certain way, swing or no swing.

Ultimately, learning to distinguish between baby chick genitalia or between strikes and balls in baseball will not help you to distinguish between different types of rocks, apples, courthouse facades, or anything else. The time spent learning how to sex chickens would not help you become a gemologist, skilled in grading jewels after a single look.

In short, we expect that becoming an expert at a particular task can allow you to operate in Perceiver Mode in that context, but only after you do an enormous amount of work. If the task is really important to you, it may make sense to invest effort in learning—but keep in mind that this is not likely to affect the mode you operate in when you are in other situations.

The Amazing Case of S.F.

In the previous section, we considered how experience can alter the bottom-brain system. A comparable story applies to the top-brain system, but in this case you can learn new strategies that allow you to use your top brain in *specific* new circumstances. This learning does not imply that you will utilize the top-brain system more in general—it just allows you to use it more in the particular task you practice.

Take the case of S.F., a student at Carnegie Mellon University who volunteered to participate in an experiment that is well known in cognitive science circles (he is identified only by his initials in order to preserve his anonymity).[7] S.F. reported to a laboratory three to five days a week over the course of more than a year and a half. At each visit, researchers K. Anders Ericsson and William Chase gave S.F. a series of random digits (for example, 4, 9, 3, 1, . . .), one every second. S.F.'s job was to repeat back each series immediately after hearing it.

In the first session, he was given a single digit and was asked to repeat it back. He was then given two random digits and asked to repeat them back; then three and so on until he could not repeat the entire sequence. On that first day, he could recall about seven digits in a row (six or seven is the usual number of random digits that people can store in short-term memory and repeat back). He returned the next day to repeat the process, and continued to return. The lists grew longer. By the end of the study, he could repeat back seventy-nine random digits!

How did S.F. accomplish this? He used top-brain processes to organize the digits into groups.

Humans can store about four groups of information in short-term memory. Within each of those groups can be up to another four groups, and so on. S.F. happened to be a long-distance runner. He was able to translate groups of digits into times for specific segments of particular races he remembered and then associate the segments. For instance, "3, 4, 9, 2" was coded as 3 minutes, 49.2 seconds. When the numbers did not fit times of familiar races, his memory plummeted; when all the digits fit, his memory was superb.

With increasing practice, he supplemented this strategy with one that coded digits into dates or people's ages. He got better and better at using these strategies to organize random digits.

Here's the kicker: After S.F learned to remember digits with such dramatic success, when he was given letters of the alphabet instead of digits, he could not recall anywhere near seventy-nine. In fact, he couldn't recall even *ten*—he was back to recalling about six items (which is within the normal range). The strategies that worked well for digits simply did not apply to letters, so he could not organize letters into groups effectively. Given time and practice, he probably could have learned new strategies—but these strategies then would apply only to letters.

In a limited number of cases, however, some transfer from a trained task to a new one is possible. Such transfer can occur if one performs tasks that share at least one underlying aspect of information processing. For example, in a study reported by Stephen and his team, participants were asked to practice "mentally rotating" objects every day for twenty-one consecutive days. As we discussed in chapter 4, mental rotation occurs when you visualize something spinning around a pivot point (either in 2-D or in 3-D), such as when you rotate a mental image of the uppercase version of the letter *p* 180 degrees to discover whether it would then be another letter. (Is it? If so, which one? Answer: Yes, a lowercase *d*.)

As expected, the participants got faster and faster at such mental rotation with practice. After three weeks, they were given another task. The researchers found that, after the participants practiced mentally rotating one set of forms, their performance in other tasks improved to a greater or lesser degree depending upon how similar the required information processing was to that required in the original task—the more similar, the better. The greatest improvement was observed when the participants mentally rotated another, similar set of forms. Participants improved somewhat when they mentally folded squares into boxes, but not as much. Participants also solved verbal analogies. The researchers found that previous practice in mental rotation did not lead to improved performance of the verbal task.

Why did participants get better at mentally folding squares into boxes after they practiced mental rotation? Because both tasks involved mentally shifting the locations of parts of objects, and this process apparently improves after practicing mental rotation—and hence it is then more effective when subsequently used in mental paper folding. Other processes required to perform mental rotation (such as keeping a figure properly aligned as it is rotated) were also practiced. Such processes apply strictly to mental rotation; that is why there was greater transfer to another rotation task than to mental paper folding. And there was no transfer to solving verbal analogies because this task shared virtually no underlying processes with mental rotation.

If we can generalize from the sorts of experimental findings we have just discussed (of which there are many) to the cognitive modes, we can infer the following: You can change the degree to which you rely on top-brain and bottom-brain functioning in a particular task or closely related tasks. Such changes could allow you to operate in Perceiver Mode (if the bottom-brain system becomes utilized more), Stimulator Mode (if the top-brain system becomes utilized more), or Mover Mode (if both the top- and the bottom-brain systems are utilized more). However, the existing scientific literature strongly suggests that the effects of such training will be limited primarily to the domain in which you practice. And you have to practice a lot to reach expert-level performance, which requires an enormous commitment. Most of us are unwilling or unable to make such a commitment for such a relatively limited reward.

In our view, no one cognitive mode is in general "superior to" or "better than" others, and there is no good reason to be dissatisfied with your dominant mode, which you'll have the opportunity to identify in chapter 13. The challenge is to find the best way to use your dominant cognitive mode to good ends, which may entail partnering with others or finding the right environment to engage your strengths. We will have a lot more to say about this, but let's first explore the four modes in greater detail, considering more thoroughly the hallmarks of each and providing illustrations.

Chapter 9

Mover Mode

In this and the next three chapters, we consider each cognitive mode in turn. At the outset, it makes sense to consider in greater detail just what we mean by a "cognitive mode."

As we've seen, you do not choose your dominant cognitive mode; your dominant cognitive mode probably is a result of your genes (most notably, those that influence your temperament) and your experience (such as the depth of your knowledge about a particular subject). If the Theory of Cognitive Modes is correct, we each have a dominant cognitive mode that typically underlies how we think and behave. Even so, in some situations we may operate differently—especially if we are very comfortable in a situation and expert in the relevant material.

According to the theory, your dominant cognitive mode is a result of how the top-brain and bottom-brain systems interact, which usually is not something you are aware of. Your dominant cognitive mode emerges from these interactions, much as an economy emerges from the interactions of individuals. The economy cannot be replaced by characteristics of individuals, and your dominant mode cannot be replaced by characteristics of the top-brain or bottom-brain system considered alone. It's how people interact that produces a particular type of economy, and it's how the two brain systems interact that produces a particular cognitive mode.

In other words, your dominant cognitive mode emerges automatically, as a result of how the top-brain and bottom-brain systems interact. However, the particular cognitive mode in which you operate at any given time is influenced by the situation you are in: In most situations, you operate in your dominant mode, but in special circumstances (if you are comfortable and expert in a particular area), you may operate in another mode.

But note that even if you do manage to operate in a different mode, you may not function well in it. The mere fact of shifting into a nondominant mode does not guarantee that you will be effective in it. Whether or not you are effective depends, in part, on how much you know about the relevant material (and hence how well you can classify and interpret the situation, using your bottom-brain system) and how well you can formulate and carry out plans (and hence how well you can respond to and anticipate unfolding events, using your top-brain system—relying on information from your bottom-brain system).

Nevertheless, it's useful to be aware not just of your dominant mode but also of the modes you may operate in when you are in particular circumstances. And it's useful to reflect on what mode other people may be operating in, which could affect how you interact with them. So we recommend that you familiarize yourself with all four cognitive modes. Don't simply take the test and then read the chapters that characterize your dominant mode.

In this and the next three chapters we present brief profiles of a contemporary person and a historical person to illustrate the nature of the modes. We are not "analyzing" these people: We cannot do that from a distance, and none of them has taken the test in chapter 13. We cannot say for sure how their top-brain and bottom-brain systems actually interact or interacted. Our purpose is not to explain what these people are or were thinking or why they actually behave or behaved as they do or did. Rather, these people offer particularly vivid illustrations of the modes in action. We draw out just those easily recognizable characteristics that allow us to clarify the nature of the modes.

Let's begin with our first character, who nicely illustrates key features of Mover Mode behavior and thinking.

A Mayor's Mayor

This was the candidate who believed he would be the next mayor of America's largest city? This man who had never run for political office and who was oratorically challenged? This unfamiliar figure who seemed bewildered by all the reporters and photographers who crowded in on him during his formal announcement at a center for the elderly in Queens, New York, on June 6, 2001?

"Would it be easier if I just wait for some pictures and get that out of the way?" the candidate said, lights flashing all around him. "That's not going to stop, no matter what I do? Well, thank you very much for coming. Ah, I'm Michael Bloomberg, and I'm running to be the one hundred and eighth mayor of the City of New York."

If ordinary New Yorkers knew of Bloomberg, it was as the billionaire who had built Bloomberg LP, the media and financial data giant. Bloomberg had no political pedigree, like a Kennedy or a Bush, nor a heroic war record or distinguished history of public service. He lacked the communication skills of a Ronald Reagan and had none of the charisma of a Bill Clinton. Business insiders knew that he had a temper and was not one for compromise. Until recently, he had been a Democrat—in a city with no shortage of high-profile Democrats who also hoped to be mayor in 2001, the last year of 9/11 hero Rudy Giuliani's tenure.

So what gave Bloomberg the confidence to run? Certainly, it helped that he had wealth, which would enable him to finance a credible campaign. But there is no shortage of wealthy people who have ventured into politics and failed.

By all appearances, Bloomberg was disposed in this context to think in Mover Mode, the mode of thinking and behaving in which people formulate and implement plans (using the top-brain system) and note the consequences of doing so (using the bottom-brain system), adjusting their plans accordingly. Their unfolding strategies typically are based on information they receive after initiating a plan, when they adjust the plan and behavior. According to the Theory of Cognitive Modes, people who are operating in Mover Mode are most comfortable in positions that allow them to plan, act, and see the consequences of their actions, and are often viewed as independent and self-directed. They often emerge as leaders, not only because they behave proactively but also because they know how to read other people and use this information when formulating and fine-tuning plans.

From his modest postwar childhood in a suburb of Boston, Bloomberg had consistently demonstrated such behavior: achieving Eagle Scout status as a young teen; excelling as an undergraduate student at Johns Hopkins University; performing well as a student at the Harvard Business School; and standing out during his early years in business, as a trader at Salomon Brothers.

We can conjecture that Bloomberg learned not just from his successes but also from his setbacks. Caught in the brutal cross fire of a leadership war inside Salomon Brothers, he was demoted after thirteen years as a highly successful trader to Salomon Brothers' tech support department—a humiliating fall from grace. But Bloomberg did not withdraw into self-pity (people thinking in Mover Mode typically are not easily discouraged): instead, he dedicated himself to a new challenge, the then-frontier of financial computing. It was that experience that led him in 1981 to found Bloomberg LP, the company that revolutionized delivery of financial information.

To New Yorkers who were learning their first real details about

Bloomberg during that June of 2001, it may have seemed that he was just another rich guy with unrealistic political ambitions and nothing but money behind them. In fact, Bloomberg had been quietly building toward this moment for many years (unlike, for example, Ross Perot—who did not plan as carefully in advance of his presidential run). Bloomberg operated in characteristic Mover Mode: He was planning, executing, registering consequences, and laying out the next moves while incorporating lessons of the preceding move.

Bloomberg first floated his mayoral ambitions in 1997, during an interview with the *Financial Times* that generated a buzz inside New York corporate circles—as Bloomberg had intended. The feedback was positive, and Bloomberg decided to change from the Democratic to the Republican party, where the opposition would be weaker, and he increased his charitable giving and participation in civic events. He schooled himself in New York's ethnic diversity. As June 2001 drew close, he hired skilled consultants—and heeded their advice. He never became a world-class speaker, but he capitalized on his strengths: intelligence, straight talk, management experience, and willingness to fight for his ambitions. He spent his money wisely and won the election.

Bloomberg was reelected to a second term in 2005 with a 20 percent margin over his opponent, the largest ever for a Republican mayor of New York. Denied the chance for a third term by a city term-limits law, he successfully campaigned for an amendment that allowed him to run again in 2009, when he won with more than 50 percent of the vote.

In his 1997 memoir, Bloomberg offered an insight into his success. "Daily, you're presented with many small and surprising opportunities. Sometimes you seize one that takes you to the top. Most, though, if valuable at all, take you only a little way. To succeed, you must string together many small incremental advances—rather than count on hitting the lottery jackpot once. . . . As a practical matter, constantly enhance your skills, put in as many hours as possible, and make tactical plans for the next few steps. Then, based on what actually occurs, look one more move ahead and adjust the plan."

It's as apt an illustration of typical Mover Mode thinking and behavior as we ourselves could have written.

The Wright Stuff

History brings us another classic illustration of thinking and behaving in Mover Mode—two people, actually, Orville and Wilbur Wright, credited with designing and building the first successful airplane.

On December 17, 1903, at Kitty Hawk, North Carolina, the Wright brothers realized one of mankind's oldest dreams: the first powered and controlled heavier-than-air flight. Less known is that these two brothers from Dayton, Ohio, achieved their breakthrough without benefit of a high school education or formal training of any kind.

The Wright brothers' fascination with flight began when their father gave them a toy helicopter, based on a design by a French aeronautical pioneer. This toy was constructed of cork, bamboo, and paper and was powered by a twisted rubber band. The boys played with it until it broke, but when it did, they were unfazed: they merely began building their own helicopters, each successive model improved with the knowledge gleaned from the previous one. Still in grammar school, the boys already exhibited behaviors characteristic of Mover Mode thinking: they embraced challenges and were not deterred by failure. Failures were not dead ends but valuable lessons in the progression to success.

The brothers' fascination with flight continued into their twenties, when German inventor Otto Lilienthal's human gliders caught their attention. After stints as self-taught printers, newspapermen, and repairers and builders of bicycles, the Wrights set out to build a glider of their own, a critical step toward powered flight.

"We knew that men had by common consent adopted human flight as the standard of impossibility," Wilbur wrote. "When a man said, 'It can't be done, a man might as well learn to fly,' he was expressing the final limit of impossibility." Refusing to accept that, the Wrights believed in themselves, a type of confidence that is frequently characteristic of Mover Mode thinking.

In 1899, the Wrights designed a double-deck glider with a 5-foot wingspan that they flew like a kite. Armed with insights gained from test flights, the brothers designed another—this one a model that could carry a man, like those built by Lilienthal. A methodical search led them to conclude that Kitty Hawk would best provide them the open space and steady winds they needed for proper experimentation, and after setting up camp there in September 1900, they arranged for the 17-foot biplane glider they had built to be shipped from their Dayton shop. There was another reason they'd chosen Kitty Hawk: Knowing that they would likely fail repeatedly before finally succeeding, they wanted soft sand to cushion the inevitable crash landings. They had taken a lesson from pioneer Lilienthal, who had died in a crash onto hard ground near Berlin.

The Wrights first flew their new glider unmanned, testing and improving their machine over the course of three weeks until they believed it was ready for a pilot to take it into the air. Wilbur volunteered and successfully completed several short controlled flights. You can see the pattern here.

Home in Dayton for the winter, the brothers built a larger, more advanced glider. In July 1901, they brought it to Kitty Hawk. The larger size had created new control problems that the brothers could not immediately solve. Using a wind tunnel back at Dayton, they spent the winter and spring experimenting with wings of different shapes. In August 1902, they were back at Kitty Hawk with a glider that had 32-foot-long wings and a tail, to stabilize the craft during turns. The months of wind-tunnel work had paid off: As 1902 drew to a close, they were ready to add a motor.

After one more winter in their Dayton shop, the Wrights returned to Kitty Hawk in September 1903 with *Flyer*, a machine featuring two propellers powered by a gas engine. The transmission and propeller shafts failed to work properly, prompting the brothers to modify their design yet again. On December 14, Wilbur took *Flyer*'s controls—but he headed too steeply into the air. The aircraft stalled and crashed into the sand almost immediately.

Three days of repairs later, it was Orville's turn. Climbing less steeply—another lesson learned—he flew about 120 feet.

"This flight lasted only about 12 seconds," Orville later wrote, "but it was nevertheless the first in the history of the world in which a machine carrying a man had raised itself by its own power into the air in full flight, had sailed forward without reduction of speed, and had finally landed at a point as high as that from which it had started."

Persisting and learning through long trial and error had paid off; success was theirs at last.

A Day in Lisa's Life

The Wright Brothers and Michael Bloomberg are examples of prominent people who illustrate what it's like to operate in Mover Mode. You might identify others who fit the description: President Franklin D. Roosevelt, who steered the country out of the Great Depression and through the Second World War; two-time Nobel Prize winner Marie Curie; Bill France Jr., who expanded NASCAR from its regional moonshining roots into the popular national sport it is today; Alvin Ailey, who created a new type of dance troupe; the Reverend Theodore Hesburgh, who brought the University of Notre Dame to international prominence during his tenure; and Oprah Winfrey, who, despite the disadvantages of a troubled childhood, built history's most successful TV talk show and a media empire.

But how does operating in Mover Mode work in the everyday world? Meet Lisa, a persona we've created: a woman in her thirties who typically thinks and acts in Mover Mode.

Lisa grew up on a farm, but during high school she decided that agriculture was not for her. She went to college, where she majored in marketing, and after some travel and a variety of entry-level jobs, she secured a position with a large technology company. She has been promoted to senior Web developer and she remains ambitious—someday, perhaps, she'd like to start her own company. She doesn't lack for friends and she's been involved with a man, Tyrone, for the last few months. They don't live together, but they see each other on weekends and often during the week. Overall, life seems pretty good.

On this Thursday, Lisa is up at 5:30 a.m., an hour earlier than usual: She has a midmorning presentation of a new website she has designed for a client and she intends to get to the office early, to rehearse one last time. But first, she has to visit the drugstore to pick up the allergy medication prescription she requested last night. After quickly dressing and eating breakfast, she walks the three blocks to the store. So far, her top-brain strength is serving her well. The day seems to be unfolding exactly as planned.

As Lisa heads to the prescription counter, she passes items she realizes she will soon need. To save herself another trip later, she picks them up now as she moves through the store: shampoo, nail polish, and sunscreen and insect repellent for the kayaking trip she and Tyrone are taking this weekend. With her prescription, the total bill comes to $42. She reaches into her pocketbook—and no wallet. Now, she remembers leaving it on her kitchen counter. (Mover Mode thinking does not ensure having a good memory!)

Lisa silently curses herself for rushing out and quickly considers her options. She knows the store manager—he'd probably extend her short-term credit—but he's not in yet. She could return her items and come back another time, but she really doesn't want to sniffle through the day without her allergy medication. She could call Tyrone, who lives nearby, but Tyrone is no morning person, and although he would certainly bail her out, he would do so grudgingly . . . and in any event, it would take him forever to get here. So Lisa apologizes to the clerk and asks him to hold her goods while she runs home to get her wallet. Returning, she reflects on a lesson for the future: Rushing carries a risk of overlooking something important.

Lisa gets to work later than she wanted, but she understands that dwelling on the delay at the drugstore could affect her presentation, so she lets it go. Similarly, although she ordinarily has another morning coffee, she knows that excess caffeine can make her jittery, so she forgoes Starbucks. Instead, after checking her overnight messages, she finds an empty room and runs through her PowerPoint presentation.

The meeting begins on time, and Lisa is ready. The presentation goes smoothly and when it's over, Lisa listens to the comments her colleagues make. She pays particular attention not only to the company president but also to the man sitting in a corner of the room, soft-spoken graphic artist Rob. Rob suggests tweaking a feature of the landing page, making the most important button noticeably larger than the others. It's a good suggestion—and Lisa, taking no offense that she didn't think of it herself, agrees to incorporate it into the final design. She never hesitates to take guidance from Rob, who frequently thinks in Perceiver Mode, or from the company president—who, like her, habitually thinks in Mover Mode. Experience has taught her to appreciate a good idea when she encounters it, regardless of its source. This is bottom-brain proficiency—recognizing good advice as good advice. The bottom brain is adept at classifying and interpreting; part of these processes involves putting new information into context, evaluating it in the broader scheme of things.

Lisa is feeling good as noon approaches and she leaves the office for lunch with John, an old college friend, now a venture capitalist, who's in town on business. For weeks, she has been looking forward to seeing him—their contact lately has been entirely by email and Facebook. She has reserved a good table at a favorite restaurant, but she needs to drive across town to get there. To her dismay, there's a traffic jam, and she realizes that she will be late. Fortunately, she thought ahead and got John's new cell phone number in their last Facebook exchange. She calls him—he's understanding—and then dials the restaurant and asks the staff to hold the reservation. (This, too, is typical Mover Mode thinking: noting glitches in a plan and revising it accordingly.) When Lisa finally arrives, she is happy to find John at their table, and the first minutes are indeed what she expected: a good conversation, sprinkled with laughter, in which they both catch up on their busy lives. But as the meal progresses, John grows subdued. For as long as Lisa has known him, he has been a steady, even-keeled man, but now it's obvious that he is troubled. Yet he does not share why.

As she observes John's body language and listens to what is now an effort to make small talk, Lisa weighs whether to say something or to pretend that nothing is wrong. With a more casual acquaintance, she thinks, pretending that everything's fine might be sensible: She could politely steer the lunch to an end, perhaps following up later with an email. But John is an old and valued friend, so she decides to speak. She chooses her words carefully, watching closely for John's reaction. Her top-brain system is formulating and executing a plan, with a particular goal in mind—and the bottom-brain system not only is interpreting what John says but also is classifying his expressions and body language in general. The bottom-brain system is passing along its assessments to the top-brain system, which uses this information to tune the plan (perhaps leading Lisa to pause before saying something mildly critical, substituting something more supportive). Top- and bottom-brain systems are working well together.

"Please tell me if I'm prying," she says, "but is something wrong?"

"No," John says unconvincingly.

"If there is," Lisa says, "you know I'd be happy to listen, and to help if I can."

"It's nothing, really," John says.

Lisa is prepared to let it go when John opens up. His teenage son has just been suspended from high school for smoking pot—and the boy now wants to drop out, in his junior year. Lisa's top-brain strengths allow her immediately to formulate several possible suggestions: motivate the boy with some reward, threaten him with some consequence, meet with the school guidance counselor and the juvenile probation officer, and more. But with her bottom-brain proficiency, she is aware of the import of John's words, facial expressions, and body language, and she reaches the conclusion that he does not want advice; he wants only someone to listen. And Lisa does, until lunch ends. They part with promises to stay in closer touch—and with Lisa's offer to be available anytime John wants her ear.

Back at the office, Lisa finds a Post-it note on her computer: The president wants to see her immediately. It turns out that a top client is demanding a revision of his company website—and he wants it this afternoon. The president believes that Lisa is the best person for the job.

"But I blocked out the afternoon for final touches on the new website," Lisa says, politely, in reference to the subject of her morning presentation.

"It will have to wait," the boss says.

Lisa finds this an unreasonable imposition: Several competent colleagues are all capable of a quick turnaround for the top client, and no one, to her knowledge, is about to sign off on a new website, as she is. Her planned afternoon will be ruined—but her bottom-brain proficiency prompts her to weigh the consequence of resisting. The owners of the new website are pretty laid-back and probably can wait until tomorrow morning, she concludes. Making a note to herself to discuss staffing at some later date with the president, she accepts the new assignment and sends an apologetic email to the owners, promising them the new design by the end of the following day. She will come in early again, just to be sure that she can get this done.

Lisa completes the new assignment by day's end. On her way home, she receives a text message from Tyrone. With a smiley face and multiple exclamation points, he is clear about his enthusiasm for seeing a particular new movie tonight. Lisa has read the reviews and seen the trailer—and has no interest. But when she calls Tyrone, she discovers that he *really* wants to see it. Lisa suggests another film they both have discussed seeing, but Tyrone—who frequently relies on Stimulator Mode—doesn't budge. Lisa's objections are not as strong as Tyrone's passion, so she accedes. Lisa knows this is a good strategy because she realizes that if she raises a stink every time she disagrees with one of Tyrone's picks, odds are that some of the time Tyrone eventually will prevail—and this will motivate him to be persistent when they disagree in the future. Lisa realizes that if she digs in her heels only when she feels very strongly, Tyrone will come to understand that she is dead serious when she demurs. Besides, because Lisa doesn't feel all that passionately about not going, she may wind up enjoying herself—if only for the discussion after the movie. And that is exactly what happens when the couple have a glass of wine at their neighborhood bar.

Later that night, as she sits in her pajamas, Lisa ponders the direction her life is taking—a reassessment she regularly makes. Of late, she has been feeling uninspired; emptiness doesn't quite describe the emotion, but she does wonder whether she is missing something. Is it her job? She is paid well and mostly feels fulfilled, but she is not in total command, as the incident with her boss today illustrated. Perhaps she needs to get more serious about starting her own company, or look to another firm that is searching for a new president. Or is it Tyrone? They certainly enjoy each other's company, but is it lasting love? Tyrone has been talking recently of living together—maybe even getting married someday.

Lisa has some soul-searching to do, but she will not act precipitously. With Tyrone, she will see how the next weeks and months unfold. As for work, she decides to begin a search through Monster.com and other sites for what opportunities might be available. After finally finishing the new website, she will call John, her venture capitalist friend, to discuss starting her own company; not only will he have sound financial advice, but he also will be able to connect her to contacts he has in the software industry. Synthesizing all this, she begins to form a plan, knowing that she can adjust it as she proceeds. She realizes it's possible that she'll conclude once again that, all factors considered, it's best to stay working where she is, at least for the short term.

Bedtime nears. Lisa knows it's time to throttle back her thoughts. She makes a cup of chamomile tea, puts some light jazz on her iPod, and soon is drowsy. She drifts off to sleep easily.

As we have emphasized, no one mode is always better than another, and none is ideal in every circumstance. According to our theory, being in Mover Mode should have obvious advantages. You are often the captain of your fate, the master of your circumstances (at least as much as they allow). If you are good at operating in this mode in a particular situation, others should turn to you as a leader in that context.

But operating in Mover Mode also has its drawbacks. If you perform inadequately, perhaps because you don't have enough experience relevant to those circumstances, you can easily offend others. Of course, if you are using Mover Mode effectively, you will adjust your behavior depending on what happens—but on-the-job learning doesn't necessarily work well in all situations.

Moreover, operating in Mover Mode can be exhausting! You need to observe your surroundings carefully, make plans, act on them, observe the results, and adjust accordingly. All of this requires energy. Sometimes it just isn't worth it. Sometimes, a cup of chamomile tea is the wise choice.

In the following chapter we consider another cognitive mode, but it, too, has its pluses and minuses.

Chapter 10

Perceiver Mode

The nineteenth-century poet Emily Dickinson illustrates well the characteristics of operating in Perceiver Mode—the mode of thinking and behaving in which people deeply engage in observing and analyzing their surroundings and circumstances (using the bottom-brain system) but tend not to implement complex or detailed plans (using the top-brain system).

Born into a well-to-do family in Amherst, Massachusetts, Dickinson was a conscientious student who had talent in music and art. She received a classical education and was exposed to the works of poets Wordsworth, Longfellow, Emerson, and Thoreau. Sickness periodically interrupted Dickinson's studies. After a few months in college, she retreated to the family home. She had no career ambitions and essentially lived day to day, occasionally entertaining friends but mostly reading and writing poems that she made little effort to have published. As the years passed, Dickinson became increasingly reclusive, although she did correspond with a number of people. Her life unfolded with little outward drama—but with a surplus of time, a gift for those who are prone to reflection.

According to the Theory of Cognitive Modes, people who habitually rely on Perceiver Mode are most comfortable when they are in positions that require them to be sensitive observers, advisers, or evaluators, and thus it was with Dickinson. It appears as if she did not utilize the top-brain system as much as she could have for devising complex and detailed plans (although she did use it, clearly, for formulating poetic narratives to describe and explain what she encountered). And when she did rely on the top-brain system for planning, such as when the day-to-day care of her chronically sick mother fell to her, she did not formulate elaborate multistep plans and did not adjust her plans based on her observations. She apparently got up every day and did what she felt required to do. Except for those duties, her time was her own.

Dickinson was a devoted gardener, and she treasured her hours with her flowers, bees, and butterflies, from which she drew insights that informed her poetry. Solitude suited her writing, but she was never a true hermit. Even at the point in her life when she rarely welcomed visitors, she kept contact with the outside world—and sought the counsel of certain trusted friends, notably writer Sarah Huntington Gilbert, who married her brother. Dickinson and Gilbert maintained a long correspondence, and Gilbert wrote the poet's obituary in 1886.

Perhaps not surprisingly, given her voracious reading, Dickinson wrote poems about the brain, which had become of great interest to nineteenth-century Americans, in part because of the publicity surrounding Phineas Gage and the great popularity of phrenology. In poem number 632 (Dickinson did not title her works), she writes lyrically of its power:

The Brain—is wider than the Sky—
For—put them side by side—
The one the other will contain
With ease—and You—beside—

The Brain is deeper than the sea—
For—hold them—Blue to Blue—
The one the other will absorb—
As Sponges—Buckets—do—

But science was not Dickinson's abiding passion; she found her greatest themes in observing nature, in the changes of season and day, in the cycles of life and death. This, too, would characterize someone for whom Perceiver Mode was the dominant way of thinking and behaving. Of the hundreds of Dickinson's poems about the natural world, we find that this one nicely captures both Dickinson's talent and the wisdom—presumably gleaned through her utilization of the bottom-brain system:

Nay—Nature is Heaven—
Nature is what we hear—
The Bobolink—the Seas—
Thunder—the Cricket—
Nay—Nature is Harmony—
Nature is what we know—
Yet have no art to say—
So impotent Our Wisdom is—
To her Simplicity.

Making Sense of the World

The Theory of Cognitive Modes leads us to expect that people whose dominant cognitive mode is Perceiver Mode often lead quiet lives, like Dickinson. Although they *do* use the top-brain system to organize information from the bottom-brain system into a coherent narrative and (at least to an extent) to plan and move ahead in life, they tend not to make or adjust complicated or detailed plans. However, because they use the bottom-brain system effectively, they tend to make sense of what they experience and they try to respond wisely. With experience, such people can become paragons of wisdom.

If the theory is correct, then people who habitually think and behave in Perceiver Mode will not ordinarily seek publicity. Still, some have achieved prominence without aggressively seeking it. History has shown that spiritual and religious figures who have helped make sense of human existence can attract large followings. Although they do not engage in self-serving campaigns, their ideas compel others.

In our own time, the Dalai Lama fits that description. Like Dickinson, he illustrates someone whose dominant mode of thinking appears to owe much to external factors (nurture, as opposed to nature): At the age of two, this son of a simple Tibetan farming family was decreed to be the reincarnation of Thubten Gyatso, the 13th Dalai Lama. Renamed Tenzin Gyatso, the 14th Dalai Lama was sent at age six to a Buddhist monastery. "The major subjects were logic, Tibetan art and culture, Sanskrit, medicine and Buddhist philosophy," he later wrote; poetry, music, and drama were among the minor subjects that rounded out his education. At the age of twenty-three, he sat for his final examination, passing with honors and receiving the equivalent of a doctoral degree in Buddhist philosophy. This achievement may have demonstrated a capacity to operate in Mover Mode, if he had to make complex and detailed plans to succeed in this particular context.

Although he describes himself as "a simple Buddhist monk," his teachings, hardly simple, transcend Buddhism. The Dalai Lama has never sought to "succeed" (in the way that someone who habitually operates in Mover Mode, such as Mayor Bloomberg, has worked to succeed), nor has he sought to change the world or even individual minds by devising plans that lead to specific actions. Rather, the Dalai Lama's thinking and behavior serve well to illustrate the results of typically relying on the bottom-brain system and allowing the bottom-brain system to drive top-brain processing (to formulate narratives). The Dalai Lama appears to try to make sense of what he has experienced, observed, read, and studied.

Our purpose is not to explore the Dalai Lama's teachings but only to note the depth of wisdom that is possible when a person habitually engages in Perceiver Mode, striving to find meaning over the course of a lifetime.

One could argue that a person who habitually thinks in Perceiver Mode is better suited to bringing a deeper perspective to human existence than is usually offered by someone who generally thinks in one of the other three modes. "It is possible to divide every kind of happiness and suffering into two main categories: mental and physical," the Dalai Lama writes in *Compassion and the Individual.*

> Of the two, it is the mind that exerts the greatest influence on most of us. Unless we are either gravely ill or deprived of basic necessities, our physical condition plays a secondary role in life. If the body is content, we virtually ignore it. The mind, however, registers every event, no matter how small. Hence we should devote our most serious efforts to bringing about mental peace.
>
> From my own limited experience, I have found that the greatest degree of inner tranquility comes from the development of love and compassion. The more we care for the happiness of others, the greater our own sense of well-being becomes.

According to the theory, everyday people who habitually rely on Perceiver Mode can contribute this sort of perspective to their families, their social relationships, and their workplaces. They needn't be the Dalai Lama.

Empathic Hannah

Perceiver Mode thinking typically does not lead to bold or dramatic strokes—or compel people who rely on it to seek credit for their accomplishments. According to our theory, people who habitually rely on this mode of thinking should generally occupy quiet corners of the world. Dickinson's genius was not recognized while she was alive. But some people who appear to engage in this mode do come to the attention of a wider audience. Along with the Dalai Lama, famed photographer Annie Leibovitz fits the bill. So do certain eminent novelists: Philip Roth, Toni Morrison, and Alice Walker among them.

Most of us don't live in the world of writers and spiritual leaders. Meet Hannah, a character we've created: a woman in her fifties who habitually thinks and acts in Perceiver Mode.

Hannah is the youngest of the six children of an artist mother and a father who was the chief policy analyst for the education department of a large state. As the youngest in her large family, Hannah learned early to observe carefully what was going on around her and act only when she felt that she had understood the situation. With parents who treasured books, she seemed naturally to gravitate to reading. A quiet and well-behaved student, she received mostly Bs and was satisfied with them, although she realized that she probably could have done better. In high school, Hannah found refuge in the library from the chaos of teenage life, so when she graduated, it was only natural that she would become a librarian. She worked as a reference librarian, fell in love with a good man, and had two children, both now in college.

Having raised her children, Hannah returned to full-time work. She has weekend duty on this particular Saturday, but the library doesn't open until noon; this leaves her plenty of time for grocery shopping. Husband Rick, an environmental lawyer, sometimes suggests that she shop online—home delivery would save time—but Hannah prefers a physical store, which allows her to stroll the aisles, contemplating new possibilities as she checks off the items on her list. Rick is golfing today and, after a leisurely breakfast, they both head their separate ways. They have discussed attending a play this evening (Rick received complimentary tickets) and the discussion will resume when they reunite, in the late afternoon.

Sure enough, Hannah makes a discovery at the grocery store: live lobsters, which the store rarely carries. After talking to the seafood manager about the best way to cook them, she buys two. Perhaps she and Rick will eat them tonight, and if not, on Sunday. Hannah completes her shopping and has moved through the checkout line when she discovers that she's missing her wallet. She puts her cart off to the side and walks to her car—but doesn't find her wallet. She vaguely remembers seeing it on the table in her bedroom and suspects that it's still there. As luck would have it, Hannah knows the manager of the store, and when she returns from her car, she asks to see her. The manager understands the situation, and she grants Hannah's request for very short-term credit. Hannah takes the groceries home, gets her wallet, returns to the store, and pays.

Hannah leaves the store to head to work, but she runs into a construction site. The street is temporarily closed while heavy equipment is demolishing a building, and cars behind her have effectively trapped her. She considers how she got into this situation and tries to understand exactly what it might mean—a temporary delay, a longer closing, a mess awaiting the intervention of police. However, she is not used to relying very much on her top-brain system to solve this sort of problem, and she does not devise strategies for breaking free, such as asking the person behind her to pull off to the side or trying to organize the cars behind her to back up. Hannah realizes that her problem could inconvenience patrons who want to get into the library, by forcing them to wait for her, and she hopes that she can get to work soon—but she waits patiently for the street to clear up. Her bottom-brain system has led her to realize that she is stuck, and her top-brain system is not prone to devising complicated plans for unraveling Gordian knots. She accepts her situation.

Hannah makes it to work on time, barely. As the library prepares to open, the director finds her at the reference desk. The director is singing a familiar tale of woe—the latest budget cuts have forced him to pink-slip the assistant whose duties included keeping the library blog. Everyone is picking up the slack, the director says, and he's assigned the blog to Hannah. Hannah listens carefully, nodding her head in understanding. No question, the blog is an important communication tool—but she understands the enormous effort required to maintain it, and she knows that she won't be paid more. That doesn't seem fair. The director senses that she is demurring. Hannah says that she will reflect on this over the next several days and perhaps come back with a compromise plan. Maybe the blog can be shared, but (using her bottom-brain system) Hannah will need to ponder more before reaching that, or any other, conclusion.

The afternoon goes smoothly—until 3:30 p.m., a half hour before closing. That's when Sally, a good friend and neighbor, stops by on the pretext of needing help with an organic diet she's developing for her nutritionist practice. The library is nearly empty, so there is time to chat. It isn't long before Hannah, who has been studiously observing her friend's demeanor, concludes that something is bothering Sally. Rather than immediately prying, she decides to allow Sally time to disclose what it is. Sally doesn't. The minutes pass in forced conversation and Hannah decides that asking is better than saying nothing.

"You seem unhappy," Hannah says.

"I am," her friend replies. "I've been diagnosed with mild depression." Hannah is not surprised: Sally, who usually seems to behave in Mover Mode, recently learned that her elderly mother is in the early stages of Alzheimer's disease.

"I'm sorry," Hannah says. "This has been so hard for you."

"Depression," Sally says. "Me, of all people."

Hannah hugs Sally. She empathizes deeply, having lost her own father last year after a succession of strokes. Sally suggests that they have lunch together next week and Hannah agrees. She will think about her friend this weekend, rereading passages from some helpful books and exploring depression on the Internet. She will call Sally to check in and almost certainly will have comfort and perspective to offer when they get together again. Hannah utilizes her bottom brain very well: Not only can she put Sally's situation in context, but she is also adept at seeing how to apply her own relevant experiences to the present situation. Her bottom brain effectively activates associated memories and primes her to behave in appropriate ways.

Rick got in his eighteen holes and is already home when Hannah pulls into the driveway. He really wants to attend the play, which everyone at work has been talking about. The reviews Hannah's read have been universally negative ("Banal is being charitable" is one that sticks in her mind). But Rick, who during weekends frequently slips into Stimulator Mode after a week of lawyerly Mover Mode behavior, is usually a source of ideas of fun things to do. Hannah, who typically does not generate many ideas for leisure activities, is happy to reflect on Rick's ideas—as wild as some of them sometimes are (this usually only entertains her). When Rick pushes to attend the play, Hannah finds herself wondering, Can any play be *that* bad? Well, maybe—and if it really is, well, she may be amused. Besides, in the grand scheme of things, two hours with the man she loves is, well . . . two hours with the man she loves. So she goes, uncomplainingly. As it turns out, the play really *is* terrible, as Rick (somewhat sheepishly) agrees. Driving home, they share some laughs.

Rick falls asleep easily at evening's end, but Hannah lies in bed awake. Her life mostly satisfies her, but Sally's mother's diagnosis has rekindled vague feelings of lack of fulfillment that surfaced toward the sad end of her father's life. Hannah will be sixty soon and will reach retirement age five years later; her children are young adults, with lives of their own. She has read and reread *Eat, Pray, Love*, and although her marriage is not an issue, she can relate to the protagonist's urge to indulge in new passions. Is she stuck in a rut? Is middle age affecting her in ways she never predicted? She has no plans to change anything suddenly. But as sleep comes, she decides to revisit this big question at another time—through reading and conversations with her husband, and with Sally and another dear friend, Maggie, a social worker.

Thinking in Perceiver Mode has distinct advantages: You can step back and get the big picture, taking your time to understand what's going on around you. If you excel at operating in this mode in a particular situation, others will soon turn to you for wise advice. If you run into trouble, however, perhaps because you don't have enough relevant experience that applies to present circumstances, you may simply have little to say.

The Theory of Cognitive Modes leads us to expect that operating in Perceiver Mode may often be personally absorbing and satisfying. You focus on understanding but are not under pressure to do something with your knowledge; you often seek knowledge for its own sake and appreciate the world around you. You sometimes live in the moment, which often is a good place to be.

However, one potential drawback of being in this mode is that it may lead you to be a bit passive. You may spend so much time in reflection that you are effectively lost in thought. This is not a necessary result of relying more on the bottom-brain system than on the top-brain system, but it is a possibility. Nevertheless, you still can use the top-brain system and move ahead—but you will tend not to have very elaborate plans.

In the next chapter we consider a cognitive mode that may underlie very different behavior, but it too has its pluses and minuses.

Chapter 11

Stimulator Mode

When people think and behave in Stimulator Mode, they often create and execute plans, utilizing the top-brain system. But because they do not utilize the bottom-brain system extensively, they do not always properly register the consequences of acting on those plans.

The Social Activist

From history comes an example of a well-known person whose life and activities we can use to illustrate Stimulator Mode: Abbie Hoffman, the antiwar activist who died in 1989. Arguably brilliant in his flamboyant way, to casual observers Hoffman may have appeared to careen through life. In his autobiography, Hoffman indeed describes being always in search of new adventures, regardless of consequence, starting with childhood.

"I was a bowling maniac," Hoffman wrote. "It was nothing for me to bowl 25 strings a day. I was a Duncan yo-yo champ at eleven. Could do a one-and-a-half somersault off a diving board and scramble like a rabbit on the basketball court. Played knock-down guts football. Was captain of my tennis team in college." After graduating from Brandeis, Hoffman was accepted by the University of California–Berkeley for postgraduate study in psychology. "My dissertation was on witchcraft," he wrote. "In experimental [psychology], while all the other students were dutifully recording galvanic skin responses and prodding rats with electric jolts, I was studying ESP under simulated stress conditions. They were expecting a potential B. F. Skinner and I was halfway down the road to Uri Geller."

After leaving Berkeley and achieving national prominence, Hoffman still often seemed, at least to casual observers, to behave erratically. But behind that public image was someone who devised detailed plans to achieve a goal—Hoffman showed every sign, in other words, that he could effectively use his top-brain system. But by all reports, he regularly failed to register the consequences of his actions and adjust his plans accordingly, and hence he did not appear to utilize his bottom-brain system as much as might have been desirable.

Consider his behavior in the protest movement that became a groundswell in 1967, when America was torn by the increasingly unpopular Vietnam War. Cofounder of the Youth International Party ("Yippies"), Hoffman organized marches, sit-ins, and demonstrations and by October 1967 was deeply involved in planning two days of actions at the Lincoln Memorial and outside the Pentagon. Preparations included obtaining a permit, which set a total limit of thirty-two hours for the demonstrations. By the time the deadline arrived, organizers had achieved their primary objective: national coverage of their cause. Many began to leave, but Hoffman and others stayed on into a second morning—and were arrested. This was pointless, given that the protest had already succeeded, and counterproductive for Hoffman, whose time would have been better spent planning the next action, not freeing himself from the criminal-justice system. With his long and intensive involvement in protests, Hoffman had repeatedly experienced the potential consequences—but he behaved like someone who did not engage in bottom-brain thinking as much as he should have.

Someone who used his bottom-brain system more extensively might have been able to classify and interpret the situation and profit from his previous experience. A key part of bottom-brain processing is accessing relevant stored memories. The mere act of classifying something allows you to access such memories, which is how you know that a blemished apple may signal that it has a worm inside it (even though you cannot see the worm) and that a cloudy day may call for bringing an umbrella. You've encountered such things before and have stored in memory what they may signal; when you encounter them again, the bottom-brain system accesses such stored memories and applies them to the current situation. Hoffman illustrates well someone who failed to benefit from such bottom-brain processing as much as he could have.

The year 1968 was a presidential election year, and Hoffman played a central role in planning major demonstrations, once again using his top-brain system. Several took place in New York City that spring, but the largest was in Chicago that August, when the Democrats were to meet for their national convention. In the weeks leading up to the convention, Hoffman oversaw production of tens of thousands of leaflets, posters, and buttons urging antiwar protesters to join him in Chicago. He helped coordinate news coverage. He reached out to speakers and musicians. He presided over weekly meetings. "Night and day we organized," he later wrote. Clearly, he could engage in intensive planning.

Hoffman's work paid off: Thousands were on hand that August 28, when the Democrats nominated Hubert Humphrey as their presidential candidate. With the world's journalists present, Hoffman had his biggest platform yet—a chance to make a powerful statement. But did he think of the potential consequences of writing the F-word in lipstick on his forehead when he dressed that morning? Probably not. But there was a consequence others might have predicted: police arrested him and detained him for thirteen hours. Hoffman missed the demonstration that would become one of the iconic protests of the 1960s—and he stood trial as one of the Chicago Seven, a long court ordeal that effectively removed him from movement leadership.

Much of the remainder of Hoffman's life was more of the same: new plans that failed to incorporate lessons that might have been drawn from previous experiences. Still, Hoffman's comments later in life are what we would expect if he shifted his dominant cognitive mode, at least on occasion and to a degree. Following a period he spent in hiding, working actively for an environmental cause, he wrote: "It's mind boggling, but being a fugitive I've seen the way normal people live and it's made me realize just how wrong I was in the past. I've grown up, too. You know how it is when you're young and not in control. I'd like to go back to school and learn how to be a credit to the community. . . . Age takes its toll, but it teaches wisdom."

In his later years, Hoffman showed signs of having developed the ability to think in Perceiver Mode at least some of the time.

Always Her Own Person

What better contemporary example could we use to illustrate the characteristics of operating in Stimulator Mode than Sarah Palin, onetime vice presidential candidate, former governor of Alaska, and continuing presence in American culture?

Whatever one might think of her politics, few would doubt that Palin moves through life by formulating and carrying out plans—but it appears that, like Hoffman, she often does not adequately register the consequences and adjust her plans accordingly. To some, the appearance of being "real" is part of her appeal. It stands in contrast to to the style of such politicians as President Obama, who often appears to operate in Mover Mode (and sometimes, to the chagrin of his supporters, in Perceiver Mode, which may incline one toward passivity).

Consider the 2008 presidential campaign, when Palin ran on the Republican ticket with Senator John McCain. Presented as a folksy, budget-cutting fiscal conservative, she commanded instant attention. Voters weary of politicians who waste taxpayer dollars applauded this governor who had pared Alaskan state construction spending, sold the gubernatorial jet, and refused to be reimbursed for her hotel stays. Given all of this, many assumed that she might favor a similar frugality with regard to her wardrobe. During the campaign, however, she and her family accepted $150,000 worth of designer outfits and accessories from Neiman Marcus, Saks Fifth Avenue, and Bloomingdale's, and indulged in an expensive makeup consultation—a spending spree that stood in stark contrast to her image as a Kmart-shopping mom. The Republican National Committee picked up the tab.

Did she not anticipate the inevitable controversy? Hadn't she learned from the experiences of other politicians who had spent lavishly on their appearance, notably Democrat John Edwards, whose well-publicized $400 haircuts just the year before helped torpedo his presidential aspirations? Faced with the fallout from this extravagance, Palin could have quickly made a midcourse correction and begun putting the controversy behind her; she could have admitted that she had made a mistake, stopped wearing the designer clothes, and moved on. But it took nearly a week of withering criticism before she even addressed "the whole clothes thing," as she called it. "I'm back to wearing my own clothes from my favorite consignment shop in Anchorage, Alaska," she said ten days before the election, but the damage had been done. And although this was certainly not the only factor, her fashion "scandal" affected her standing in the polls, as she dropped ten points in perceived leadership abilities, according to a CNN/Opinion Research survey.[1]

Consider another Palin controversy. In March 2010, she posted on her Facebook page pictures of gun cross hairs that "targeted" Democratic members of Congress for defeat (ostensibly for their support of President Obama's health care overhaul). Several who were singled out, including Rep. Gabrielle Giffords of Arizona, had already received death threats or been victims of vandalism. Someone operating in Mover or Perceiver mode probably would have foreseen that this posting could come back to haunt her, but Palin's cross hairs were still online on January 8, 2011, when Giffords, who herself had vociferously objected to the posting, was tragically shot. And someone operating in Mover or Perceiver mode would probably have promptly acknowledged a mistake and apologized, but Palin did not. Instead, five days later, as Gifford lay in a medically induced coma, her survival still in question, Palin posted a video in which she castigated the media for their coverage of the shooting.

"Especially within hours of a tragedy unfolding, journalists and pundits should not manufacture a blood libel that serves only to incite the very hatred and violence they purport to condemn," Palin said.[2] Once again, she apparently had failed to register the consequences of her actions. Many took umbrage at her use of the term "blood libel," a slur that has historically been used against Jews (Giffords is Jewish). Rather than deflating the cross hairs controversy, Palin only compounded it.

Regardless of Palin's intent—whether she wishes to be controversial or only intends to act in accordance with her beliefs—she frequently behaves like a person operating in Stimulator Mode.

"I am the first to say, 'buck up or stay in the truck,'" Palin wrote in *Going Rogue*, her autobiography. "You have a choice about how to react to circumstances."

Indeed, people do. And choices can be influenced by previous consequences, but it appears that Palin has frequently failed to assimilate those. She closed her autobiography with political advice that incorporates the need for making and implementing plans—but leaves unanswered the issue of how to respond to the results of acting on such plans, when consequences might be unfavorable.

"The way forward is to stand and fight," Palin wrote. "March on Capitol Hill. Write letters to the editor. Run for local office—you never know where it may lead."

That may be true, but in Stimulator Mode, you often don't consider how to revise your plans if they go off the rails. And sometimes that can get in the way of moving ahead.

Andy, the Radio Guy

Many other prominent people could be used to illustrate the characteristics of thinking in Stimulator Mode: the singer/songwriter Courtney Love, who has led her bands to success many times but consistently veers out of control; Tiger Woods, who often appears to operate in decisive Mover Mode on the golf course but not always in his personal life; talk-show host Glenn Beck, who has not changed his obsession with conspiracy theorizing despite negative consequences (including a decline in ratings that prompted Fox News to drop him). On the other hand, Stephen Colbert has used the appearance of operating in Stimulator Mode to his advantage, and many creative types appear to benefit from it.

Bringing this back from the celebrity world, meet Andy, a character we've created: a man of about forty who habitually thinks and acts in Stimulator Mode. We use him as a way to make vivid the nature of Stimulator Mode, as well as its pitfalls and advantages.

Andy was just three when his parents divorced. His mother raised her only child, waiting tables and singing part-time with a band. Andy was a good student, but school bored him. He was twelve when he picked up a guitar, fourteen when he formed his first garage band, and seventeen when he finished high school. His mother insisted that he go to the state university. He did—and hated it. There was no real outlet for his music, but he did join the school radio station and he fell in love with broadcasting. When the manager of a small commercial station offered him a job, he quit college. He's worked at rock stations ever since.

On this Monday, the alarm wakes Andy at 6:00 a.m., the usual time. He runs through his plans for the day as he makes coffee and leaves a good-morning note for the still-sleeping Pam, his girlfriend, who recently moved in with him. His immediate issue is getting to Target, where he will buy everything he needs for the surprise birthday party he is throwing for the noon-to-three deejay at Classic Rock 78.8, where Andy's been program manager for six years. Andy fills his cart with balloons, noisemakers, hats, paper plates, plastic utensils, and everything else he needs. Last stop is the freezer section, which holds ice cream cakes, the deejay's favorite (she talks incessantly about her weakness for ice cream cakes). But Target is out of stock. This wasn't part of the plan. Andy will have to go with a regular cake, but he's not happy.

When he gets to checkout, he is embarrassed to discover that he cannot remember the password for his Classic Rock 78.8 credit card. He is carrying only about $10 in cash, and no one answers when he calls the station for the password. A line is building behind him, but Andy barely notices. He's ready to leave everything and somehow deal with the party later (he's not sure how exactly, but now's not the time to worry) when the clerk suggests that perhaps he has another credit card. Andy does have his personal card and he presents it, reluctantly: the station is notoriously slow in reimbursement. But it *is* a solution, albeit not of his making. Although his bottom-brain system did not help his top-brain system to devise a new plan when he ran into the glitch, his top brain is engaged and he returns to his plan for the day.

How was Andy's top-brain system able to formulate and carry out these plans? It operated by analyzing the current situation. The top-brain system is adept at solving problems—but it does so only if the person recognizes that a problem exists and then thinks through the steps necessary to solve the problem. This is a slow, deliberate, and effortful process. In contrast, the bottom-brain system can classify and interpret situations, automatically triggering relevant information stored in memory—which often can solve a problem before a person is even consciously aware that it exists. In Andy's case, he recognized the problem and formulated a way to address it. This required a lot of effort, and Andy would have been better off if his bottom-brain system had fed the relevant information to his top brain. But this was not the case.

Leaving Target, Andy heads to the radio station—and soon gets stuck in a horrible traffic jam. He looks for a way to get through the traffic, trying to drive on the shoulder, leaning on his horn—in hopes that the automotive waters will part before him. No such luck.

After he finally arrives at the station, things only get worse when boss Jack finds Andy in the lunchroom, where he is unpacking the party goods. The Arbitron ratings that came in earlier this week showed increases in two closely watched numbers: the all-important average quarter-hour share and the cumulative audience, but is Jack considering Andy's role in that? No. Jack is intent on a rebuke.

"I need to talk to you," Jack says.

"About what?"

"You need to schmooze our top accounts more. I've told you that repeatedly."

"And I've told you repeatedly I hate suits."

"I'm not asking you, Andy," Jack says. "I'm ordering you."

"And are you paying me more?"

"You know the answer to that."

"And you know my answer to your answer. It's not part of my job. End of story." Andy goes to his cubicle to check the overnight emails and messages.

Andy's bottom-brain system was not functioning in a way that would have allowed him to reflect on the situation and see the tradeoffs between the short-term gratification from being a wiseass and the possible negative long-term consequences for his relationship with Jack. In particular, his bottom brain has not learned to recognize the sorts of situations in which emotional impulses need to be inhibited—and hence does not send that information to the top brain, which has the job of inhibiting such impulses. It is the top brain, in most cases (specifically, the front parts of the top brain), that inhibits other parts of the brain—but only if it receives appropriate information that will lead it to formulate and carry out such processing.

Fortunately, both Andy and Jack have moved past their confrontation by 10:30 a.m., when the staff begins gathering in the lunchroom for the party. Deejay Shannon is always punctual, arriving an hour before airtime to get ready—but 11:00 a.m. comes and goes, and soon it's half past the hour and no Shannon. With ten minutes to the start of her show, she walks in; her car wouldn't start, so she had to call a taxi—which took almost an hour to get her. So much for the party. Shannon goes straight to Studio B. Andy is a bit disappointed and annoyed. Only when one of the secretaries suggests they improvise with what she calls a "rolling party"—spread through Shannon's off-air breaks and ending when she signs off—does his mood improve.

Quickly adjusting to changing circumstances is not Andy's strong suit (like others who habitually rely on Stimulator Mode, he can formulate plans easily but isn't good at using new information to revise those plans). Colleagues could remind him of a stunt he engineered last year. He hired young women in string bikinis to sing the Classic Rock 78.8 jingle during afternoon drive time. Standing on a flatbed truck parked on an interstate overpass, they indeed produced the intended result, mass attention—until a car crashed, tying up traffic for miles. Andy did not respond well to this unintended consequence: he kept the girls singing and the remote broadcast going, even after he could hear the sounds of approaching sirens and a police helicopter circled. Andy was briefly arrested, and when the mayor asked the FCC to suspend the station's license, Jack considered firing him. The negative press overshadowed the momentary rating boost.

An afternoon of meetings unfolds without crisis, leaving Andy pleased, if bored by the same old discussions. When a good friend,

Mark, calls and asks whether Andy can sneak out for coffee at a nearby café, he welcomes the break. Arriving at the café, Andy fails to notice that his friend is troubled. He talks merrily for ten minutes before picking up on the fact that Mark has been stone-silent and staring down at the table the entire time. It finally dawns on him that all is not well with his friend.

"You okay, man?" Andy asks.

"Never been better," Mark says.

"You sure?"

"Just tired is all," Mark says.

Andy doesn't probe further. But a few minutes later, Mark discloses that he was not offered a job for which he had campaigned aggressively. Andy apologizes for not remembering that he was pursuing the opportunity. But it doesn't occur to him to apologize for not having noticed his friend's mental state earlier. Andy's bottom brain simply does not classify and interpret social cues well, and the top brain does not receive the information that would have led it to initiate such a response.

When Andy gets home, Pam, a music writer and blogger, makes a quick dinner. She has passes to a concert by a new local band, Devoid of Reason, but Andy doesn't like them—he could not get through their demo MP3 file, wholeheartedly agreeing with the state's major newspaper's music critic when he wrote that their sound "is devoid of everything, starting with appeal." Pam says the sound just isn't for him, and who does he think he is to pass judgment anyway? They've had this sort of discussion before and it's led to arguments—like Andy, Pam habitually thinks in Stimulator Mode. But this time, remembering how much he values their relationship, Andy gives his girlfriend the benefit of the doubt. They attend the concert.

Andy likes Devoid of Reason even less live than he did when he heard their demo, but he keeps his opinion to himself (mostly). Unable to get into the music, he lets his mind drift away. With the exception of Pam, things lately have been feeling stale, as if he's seen this movie one too many times before. He's lasted six years at Classic Rock 78.8, an eternity in his business—and too often he's bored. He just needs a change; it's really that simple. Tomorrow, he'll start working his connections to see whether there's something better out there.

And this, more than the concert, is what Andy and Pam discuss back at the apartment. She's up for a change, too—and willing to roll the dice and see what happens. But if the consequences of their plans (such as moving to another city) aren't as anticipated—if their new apartment turned out to be too noisy, for example—neither of them will quickly develop a solution to the problem.

In spite of certain drawbacks, being in Stimulator Mode has some advantages. For instance, you can generate plans and stick to them, allowing creative ideas to come to fruition (Steve Jobs sometimes seemed to operate in Stimulator Mode). And if you are good at operating in this mode in a given situation, others may often turn to you as a source of ideas. Moreover, being in Stimulator Mode offers a degree of freedom, of not being nudged by small things that happen around you.

Still, if our theory is correct, then the downside is clear: People who are operating in Stimulator Mode can be bulls in a china shop and can easily offend others. In Stimulator Mode, you often may not adjust your behavior in response to the results of your plan—and that can be a problem if your plan turns out to have been inappropriate for the circumstances. In addition, being in Stimulator Mode can be frustrating. You can make good plans and still be blindsided. This can lead to feeling unappreciated by others, misunderstood, and dissed.

In the next chapter we consider the fourth mode of thinking, which may require the least effort of any cognitive mode.

Chapter 12

Adaptor Mode

Like people who usually operate in Perceiver Mode, people who habitually operate in Adaptor Mode typically lead lives away from the limelight. By definition, they do not usually devise and carry out complex or detailed plans (using the top-brain system); nor do they usually interpret and understand in depth what is happening around them (using the bottom-brain system, which sometimes also recruits the top-brain system to formulate explanatory narratives). They thus can easily be pushed and pulled by immediate events.

But this characterization also implies that people who habitually rely on Adaptor Mode will easily adapt to plans set by someone else: they can be valuable team players. According to our theory, people who habitually rely on this mode in business should often function as the essential infrastructure of the organization, handling the critical operations of the enterprise. And people who habitually rely on Adaptor Mode should often be relaxed and easy to be with; they don't worry too much about what they should do in the future or what they should have done in the past. We would expect them often to be great fun.

We can find no better illustration of the characteristics of operating habitually in this mode than the behavior of Yankees star Alex Rodriguez, widely known simply as A-Rod, whose talent and long tenure in New York, media capital of the world, put him on a grand stage. Given his fondness for partying and pretty women, Rodriguez cannot be accused of failing to enjoy life.

Rodriguez came to the Yankees in 2004 after record-breaking seasons with the Texas Rangers. A two-time Gold Glove shortstop, he agreed to switch to third base to accommodate new teammate Derek Jeter, Yankee captain and another baseball great. Jeter had been the Yankees starting shortstop since 1996. "I don't see it as a big deal at all," Rodriguez said. "I've always thought of myself as a team player. Playing third base is the ultimate team move." The years passed and A-Rod stayed at third, to some extent in Jeter's shadow, largely without complaint. For the most part, he registered what was going on around him and behaved in accordance with his perception of what he was expected to do. He was indeed a model team player, most of the time—no easy feat on a legendary team long filled with outsize egos and Hall of Fame talent.

But off the field, Rodriguez's behavior brought repeated troubles. Despite marrying and having two children, he continued to see other women—publicly. Had he somehow failed to understand the New York tabloids? Did he not heed his own advice—"It's about life in general, *managing* life" (emphasis added)—when he told an audience in 2005 that he had sought therapy? When his name, sometimes with the nickname "Stray-Rod," began surfacing in stories linking him to strippers and a fitness model, he apparently did not foresee the consequences and continued his behaviors as the tabloids continued to feed. Reports of romantic involvement with pop singer Madonna in the summer of 2008 were the last straw for his wife. She filed for divorce.

Rodriguez is hardly the first celebrity who has failed to grasp the negative implications of repeated publicized indiscretions. A more telling illustration of possible consequences of operating in Adaptor Mode may be found in his use of steroids.

By Rodriguez's time, no one in baseball could be ignorant of the steroids scandal. The public roots date to 1988, when the *Washington Post* wrote about Jose Canseco's use of the performance-enhancing drugs. Congress passed legislation creating criminal penalties for the illegal use of anabolic steroids, but press reports of baseball players who took them continued. The reports increased during the record home-run seasons of the aging Mark McGwire, who hit 70 home runs in 1998 and later admitted to using steroids; and Barry Bonds, who hit 71 home runs in 2001 and was soon embroiled in the scandal for his alleged use. In 2003, Major League Baseball began testing for the substances, which meant that potential evidence of criminal activity would be collected. And in 2005, the name of pitching star Roger Clemens publicly surfaced in the steroid scandal for his alleged use, after years of rumors inside the baseball world.

Surely amid all this, a player bound for baseball immortality would not want to use the substances. Rodriguez repeatedly denied the rumors until confronted with an airtight *Sports Illustrated* story in 2009, at which point he confessed.

Still, Rodriguez apparently had not learned the full lesson: In February 2010, the *New York Times* reported that he had received treatment from a Canadian sports doctor who was under investigation for giving athletes human growth hormone. The doctor confirmed that he had traveled to New York to treat the star but claimed that the treatment involved only anti-inflammatory medication for an injury. But the damage was done, and more was coming. A-Rod could have avoided contact with a cousin who allegedly provided him with steroids prior to 2009—a cousin who was banned by the Yankees from any team-related activity or nonpublic facility. But Rodriguez was spotted with the cousin again in spring 2011.

With the passage of years, however, Rodriguez shows signs of having come to rely more on bottom-brain processing. He has invested in art, become a generous philanthropist, and established himself as a champion of Boys & Girls Clubs of America. He wrote a children's book. All this may illustrate newfound Perceiver Mode thinking and behaving. It suggests that—with appropriate effort and experience—no one is necessarily locked for life into one cognitive mode in all circumstances.

A Consummate Actress

Like Rodriguez, Elizabeth Taylor was blessed with natural talent. She also possessed rare beauty. The combination brought her great commercial and critical success, but it hardly followed a detailed plan: like A-Rod, Taylor often was carried along on the currents of circumstance. She managed her brilliant acting career—four Golden Globes, two Oscars—and that suggests an ability to operate in Mover Mode at least part of the time. But when it came to personal relationships, Taylor behaved, from a young age, as if she regularly operated in Adaptor Mode.

Given his reputation as an obnoxious and abusive drunk, hotel heir Conrad Hilton, previously married (to Zsa Zsa Gabor), would have seemed a poor choice as first husband for Taylor, then eighteen. Hilton was given to extreme mood shifts and was a notorious womanizer, dubbed "the man with 100,000 beds." He married Taylor in 1950—and in January 1951, less than a year later, he became Taylor's ex-husband number one. The divorce seemed to suggest that although Taylor had made a mistake—perhaps understandable given her youth—she had learned from the experience.

Apparently she had not. After dating several men, Taylor settled on Michael Wilding—an English actor who had been married before and who was subject to dramatic shifts of mood. Was this not familiar? A year after divorcing Hilton, Taylor announced her engagement to Wilding. They married in February 1952. The Wildings had two children, but, dissatisfied with her second husband, Taylor began seeing other men. Wilding created his own scandals, and Taylor was done with him—and on her way to the twice-married producer Michael Todd, whose volatile temper was legendary. Their rocky relationship ended in 1958, when Todd was killed in a plane crash. Taylor was soon seeing singer Eddie Fisher, who was still married to Debbie Reynolds. After Fisher's divorce, Taylor married him, in 1959. Then, on the set of the movie *Cleopatra*, released in 1963, Taylor became involved with Richard Burton.

"Since I was a little girl," she wrote in a memoir, "I believed I was a child of destiny, and if that's true, Richard Burton was surely my fate." This attitude captures well the kind of thinking that occurs in Adaptor Mode, for it suggests that she felt that she was just carried along by external events, not learning from her experiences and not devising and following detailed or complex plans. But such behavior can carry an unintended price, as someone thinking in Perceiver Mode would probably realize.

Burton was an alcoholic, philanderer, and abuser—the worst qualities of Taylor's previous husbands all rolled into one. And yet, like Taylor, he could be romantic and fun. They married in 1964, and their marriage became what others had predicted: tempestuous and booze-filled, a real-life version of *Who's Afraid of Virginia Woolf?*—the 1966 film in which they starred and which brought Taylor her second Oscar. By 1973, Taylor had had enough. She separated from Burton and they divorced the next year, but in a press release, Taylor gave little evidence that her head had finally overruled her heart: "I believe with all my heart that the separation will ultimately bring us back to where we should be—and that's together!"

They did indeed get back together, and, in October 1975, they walked the aisle again. A predictable pattern recurred: Burton drank, Taylor battled her own demons, and the couple fought. In 1976, Taylor left Burton for the last time. We will gloss over Taylor's next two marriages, both of which ended in divorce but which apparently brought her a degree of happiness—though not necessarily late-life wisdom. Asked once why she had married so many times, Taylor said: "I don't know, honey. It sure beats the hell out of me." The fact that she didn't analyze events that she repeatedly experienced is characteristic of someone who infrequently relied on (the optional, nonreflexive types of) bottom-brain functioning. In her personal relationships, Taylor seemed to personify Santayana's famous observation, "Those who cannot remember the past are condemned to repeat it."

But as we have stressed, a habitual dominant mode of thinking has nothing to do with goodness, emotion, or intelligence; nor does it mean someone is forever trapped and forced to think in that mode, stuck in one mode in every arena of life and incapable of shifting. Taylor illustrates such a shift well: She converted from the Christian Science faith of her childhood to Judaism, the result of a spiritual journey that demonstrated a capacity for Perceiver Mode thinking. She found success as a jewelry designer and creator of perfumes, becoming something of an entrepreneur—as if she sometimes thought in Mover Mode. She became a respected humanitarian and philanthropist, raising awareness and hundreds of millions of dollars for HIV and AIDS-related illness, achievements that also characterize thinking in Mover Mode.

A Young Man Named Nick

Although thinking in Adaptor Mode is not likely to attract the limelight, some people who show signs that they habitually think in this way, like Taylor and A-Rod, do get wide attention. Britney Spears and former Minnesota Governor Jesse Ventura are among the public figures who show evidence of often operating in Adaptor Mode. So did baseball great Mickey Mantle, whose off-field behaviors (complicated by alcoholism) proved hurtful to his family and friends—although later in life, Mantle illustrated Perceiver Mode thinking when he expressed remorse and tried to make amends.

Let's return to the everyday world. Meet Nick, a character we've created: a man in his late twenties we've designed to illustrate what it means to think and act in Adaptor Mode.

Nick is the older son of parents who raised their two children in the suburb where they themselves were born. Nick's mother worked as a paralegal, while his father was an electrician and shop steward. Nick was an easygoing child who played in the Little League, joined the Cub Scouts, and never gave a thought to college: starting when he was a young teenager, he apprenticed with his father, who took weekend jobs to supplement his income. After high school, Nick promptly found work as a journeyman electrician, and his hard work paid off when he passed his master electrician's exam when he was just twenty-three. He works now on a construction crew that is building a federal courthouse complex, a year-long project.

On this Friday, Nick wakes at dawn with his wife, Erica, a stay-at-home mom to their three young children. The children still sleep. Nick and Erica eat breakfast and Nick makes the children's lunches and leaves a note for each. Beyond planning to buy a new power drill on his way to work, he is not thinking much about the day ahead; in all likelihood, it will unfold as most days do, according to the foreman's directions. On his way out, he kisses Erica and tells her again how lucky he is to have her. Erica often operates in Mover Mode, and it's good for the family that she does: She manages the household budget, does most of the shopping, and keeps the kids on track in school and in their many afterschool activities. Her dominant cognitive mode nicely complements her husband's—a fact that she, although not necessarily he, may have intuited when they decided to marry and raise a family.

Nick finds the Makita drill he wants at Home Depot, which—luckily for him—opens at 6:00 A.M. Picking up a set of drill bits, he proceeds to a self-checkout station, where he slides his work-expense debit card, scans the items, and waits for approval from Visa. The charge is declined—and only now does he remember he neglected to ask Erica to transfer funds from their main bank account (it was her idea to have a separate work account, to simplify taxes). By now, a clerk is on the scene. Embarrassed, Nick mumbles an apology, returns the items, and leaves. He could have called Erica and asked her to transfer the funds online now, but that didn't occur to him; in any event, he still has his old drill, and although it's lost some power, it will get him through another day. Writing himself a note to remind Erica to transfer the money when he gets home, he heads off to the courthouse project. The incident at Home Depot fades from his mind.

A short distance from work, Nick becomes stuck in traffic. He's frustrated, but as soon as he realizes that there's really nothing he can do, he relaxes and listens to his iPod. If he thinks of it, he might phone in to let his foreman know that he is stuck—but he's into his music, and it doesn't occur to him. His bottom-brain system does not lead him to see broader implications of his current situation (its effects on other people, such as his foreman), nor does he take advantage of the time to use his top-brain system to make plans about things that really matter to him. Instead, the immediate situation is driving his agenda, as we expect is typical of people who are operating in Adaptor Mode. His top brain is not formulating complex or detailed plans that would guide his thoughts or behavior; instead, he waits for external guidance about what to do next.

Fortunately, Nick arrives on time. He goes to the coffee truck, where he finds Jake, a high school buddy who took the same vocational courses and followed him into the construction trades, becoming a crane operator. They met in middle school and have been close ever since. More than once, Nick has sought advice from Jake, a man of quiet assurance (who seems typically to think in Perceiver Mode and always gives thoughtful answers). The two men recap last night's basketball game, but Jake, a die-hard fan, lacks enthusiasm. After several dead spots in the conversation, Nick realizes that his friend is uncharacteristically gloomy. Still, Nick says nothing . . . until Jake drifts off in midsentence, his thoughts unquestionably elsewhere.

"So, what's up?" Nick says.

"What do you mean?" Jake says. He knows what Nick means.

"It's Friday, the weekend, and you're acting like you've been to a wake. You okay? Did you have a fight with your wife or something?"

"No," Jake says, "things are fine."

"You sure?"

"I'm sure," Jake says.

And with that, Nick moves along. It's 7:00 A.M., start of the workday.

The foreman reviews today's plan with his crew, and the electricians head to their stations—all but Nick. The foreman asks him to stay behind. He has a special assignment for the young master electrician: all next week, he wants him to take a new apprentice under his wing. Nick knows what this will entail: essentially, babysitting. Nick certainly understands the need for apprenticeship, having been there himself, but why did the foreman have to choose him to mentor the kid? There are plenty of other electricians with more experience who could handle the job.

But the foreman has not asked Nick; he's ordered him. And although Nick might win the battle if he pushed back hard (the foreman values him as one of his best workers), he decides it's not worth it: The order isn't *totally* unreasonable, and besides, good relations with the boss count for a lot. So Nick makes a joke about babysitting, which draws laughter from the foreman, and agrees to take on the apprentice next week. He'll even treat him to coffee.

The contractor always knocks off early on Fridays, and Nick is home by 3:30 p.m. He plays with the children, then helps Erica prepare the family dinner. As the family eats, talk turns to Friday night, which traditionally has been Date Night for Nick and his wife. Erica has arranged for her mother to watch the kids. She wants to see the romantic comedy that's opening this weekend.

Nick enjoys movies, but he usually doesn't like romantic comedies. So he good-naturedly suggests they see something else—maybe compromise on the latest Johnny Depp film. But Erica is adamant, and Nick graciously agrees. He is just going with the flow; this is often characteristic of Adaptor Mode thinking. Adaptor Mode thinking doesn't lead a person to generate complex or detailed plans that guide thoughts and behavior; instead, other people or events serve to steer the person. And although he may not articulate it quite this way, his acquiescence also is a gesture of appreciation for the workload Erica carries, mostly uncomplainingly, with their family.

After the movie, Erica and Nick stop by Applebee's for a beer. They chat a bit about the film and then talk about the community college night courses that Erica is soon to start, as a first step in her plan to become a nurse, a career she wants when the children are older. Nick is supportive, and the conversation prompts him to examine his own career, not a regular exercise for him. Still, from time to time, he *does* wonder if he wants to string wire forever—if maybe he'd like to move up the union ladder or try something else entirely, like pursue his childhood dream of becoming a firefighter. He's discussed his dream before with Erica, and when he broaches the subject again this evening, she gives the same sound advice: enroll in an EMT course, join a volunteer fire company, apply for fire academy. It would be difficult, she says, but he could probably manage while still keeping his day job and remaining a good dad.

But Nick senses that chasing his old dream would require detailed, long-range planning. The planning would be likely to require adjustment as circumstances changed, as almost inevitably they would (as Erica has rightly observed). Right now, it seems too much to undertake. Overall, life is pretty good as it is. Why rock the boat?

Being in Adaptor Mode has some clear advantages. When you relax, you really relax—you don't fret about the future or obsess about the past. Moreover, because you very likely are easy to get along with in this mode, other people often enjoy your company. The downside, according to the Theory of Cognitive Modes, is that you will often be buffeted by the world around you—and that can be detrimental. As psychologists showed long ago, animals that have some control over their environment experience less stress (and fewer ulcers) than animals that are always on the receiving end, having no such control.[1]

Chapter 13

Test Yourself

Now that we've reviewed each of the cognitive modes in detail, you may have some ideas about your own dominant mode of thinking—the mode that you are most comfortable operating in most of the time. The test we present in this chapter will allow you to be more definitive. Although you may not always rely on the same mode in every context, people's responses to the test indicate that they do operate in a single mode most of the time. So get a pencil and paper and determine yours—or take it online at www.TopBrainBottomBrain.com and have your score computed automatically.

This test is called the Dorsal-Ventral Questionnaire (recall that dorsal means "back" but refers to the "top" in two-legged creatures; and ventral means "bottom"). It was developed by Stephen and his longtime collaborator William L. Thompson.[1]

For each statement in the test, use the following scale:

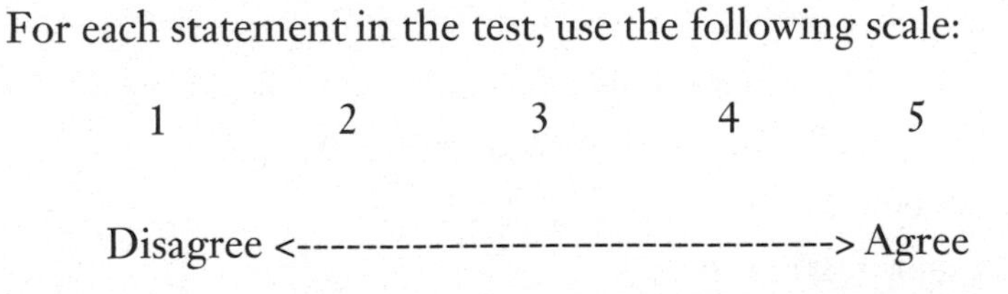

where 1 indicates that you completely disagree with the statement; 2 indicates that you somewhat disagree; 3 indicates that you neither disagree nor agree; 4 indicates that you somewhat agree; and 5 indicates that you completely agree.

1. When I look at a garden, I usually notice the patterns of plantings.
2. If I like a piece of furniture, I want to know exactly where it will fit in my home before I will buy it.
3. I prefer to make plans about what to do before I jump into a situation.
4. In a museum, I like to classify paintings according to their style.
5. I try to examine items in a store very carefully.
6. I like to assemble all the necessary tools before I begin a project.
7. I prefer to call ahead to a hotel if I may not get there until late in the day.
8. As a rule, I try to react appropriately to my environment.
9. I like to examine the surfaces of objects in detail.
10. When I first turn on the TV, I like to identify specific people on the screen.
11. I effortlessly note the types of dogs that I see.
12. I like to think about what to expect after I make a decision.
13. I like to look at people's faces and try to classify where their ancestors came from.
14. I think of myself as someone who plans ahead.
15. Before I buy a new shirt, I think about whether it will go with my other clothes.
16. When I hear music, I like to identify different instruments.
17. I take the time to appreciate paintings when I go to an art exhibition.
18. I enjoy making plans.

19. In the morning, I often think ahead to what I'll need to do that day.
20. I prefer to examine objects closely enough to see how color changes on their surfaces.

After giving the test to hundreds of people, the researchers calculated two sorts of measures that are crucial for scoring it and interpreting what the scores mean:

One measure was the mean (that is, average) values for all of the ratings for top-brain items, which was 37 (out of a maximum of 50), and the mean values for all of the ratings for bottom-brain items, which was 33 (out of a maximum of 50). The other measure was the standard deviation for each type of rating, which indicates the spread of the ratings above and below the mean. The standard deviation ended up being 6.4 for both scales.

To get your score:

First, add up your ratings for items 2, 3, 6, 7, 8, 12, 14, 15, 18, and 19. This is your top-brain score.

Second, add up your ratings for items 1, 4, 5, 9, 10, 11, 13, 16, 17, and 20. This is your bottom-brain score.

As is clear from the scoring, the test is not all-or-none, and hence there are gradations in the degree to which you may habitually operate in one mode or another.

Top-Brain Scores

Let's start with the top-brain ratings, and consider scores above the average (or mean)—which indicate habitual use of top-brain processing in optional ways (when you are not required to do so by the situations), to one degree or another. Specifically, if your score was over 47 (about 1.5 standard deviations above the mean), you have a *very strong tendency* to use top-brain processing in such ways; if your score was over 37 but below 47, you have a *tendency* to use top-brain processing in such ways. In what follows, we will be discussing only such optional ways of using the brain systems.

Now let's consider top-brain ratings that are below the mean of 37. These scores indicate that you do not habitually use top-brain processing. Specifically, if your score was under 27 (about 1.5 standard deviations below the mean), you have a *very strong tendency* not to use top-brain processing; if your score was under 37 but greater than 27, you have a *tendency* not to use top-brain processing.

TOP-BRAIN PROCESSING

Over 47	Very strong tendency to use top-brain processing
37–47	Tendency to use top-brain processing
27–37	Tendency not to use top-brain processing
Under 27	Very strong tendency not to use top-brain processing

Bottom-Brain Scores

Now consider your bottom-brain ratings. If your score was over 43 (about 1.5 standard deviations above the mean), you have a *very strong tendency* to use bottom-brain processing (in optional ways, when you are not required to do so by the situation); if your score was over 33 but below 43, you have a *tendency* to use bottom-brain processing.

Finally, let's turn to bottom-brain ratings that are below the mean of 33, which indicate that you do not habitually use bottom-brain processing. Specifically, if your score was under 23 (about 1.5 standard deviations below the mean), you have a *very strong tendency* not to use bottom-brain processing; if your score was under 33 but greater than 23, you have a *tendency* not to use bottom-brain processing.

BOTTOM-BRAIN PROCESSING

Over 43	Very strong tendency to use bottom-brain processing
33–43	Tendency to use bottom-brain processing

23–33	Tendency not to use bottom-brain processing
Under 23	Very strong tendency not to use bottom-brain processing

Now, the last step: To identify your dominant cognitive mode, look up your top-brain and bottom-brain classifications in the following table.

TOP-BRAIN/BOTTOM-BRAIN CLASSIFICATION	**TYPICAL MODE**
Very Strong Tendency Top/ Very Strong Tendency Bottom	Consistently Mover Mode
Tendency Top/Tendency Bottom	Mover Mode, but particularly context-dependent
Very Strong Tendency Top/ Tendency Bottom	Mover Mode, but sometimes Stimulator Mode
Tendency Top/ Very Strong Tendency Bottom	Mover Mode, but sometimes Perceiver Mode
Very Strong Tendency Not Top/ Very Strong Tendency Bottom	Consistently Perceiver Mode
Tendency Not Top/Tendency Bottom	Perceiver Mode, but particularly context-dependent
Tendency Not Top/ Very Strong Tendency Bottom	Perceiver Mode, but sometimes Mover Mode
Very Strong Tendency Not Top/ Tendency Bottom	Perceiver Mode, but sometimes Adaptor Mode
Very Strong Tendency Top/ Very Strong Tendency Not Bottom	Consistently Stimulator Mode
Tendency Top/Tendency Not Bottom	Stimulator Mode, but particularly context-dependent
Very Strong Tendency Top/ Tendency Not Bottom	Stimulator Mode, but sometimes Mover Mode
Tendency Top/ Very Strong Tendency Not Bottom	Stimulator Mode, but sometimes Adaptor Mode
Very Strong Tendency Not Top/ Very Strong Tendency Not Bottom	Consistently Adaptor Mode
Tendency Not Top/ Tendency Not Bottom	Adaptor Mode, but particularly context-dependent

Very Strong Tendency Not Top/ Tendency Not Bottom	Adaptor Mode, but sometimes Perceiver Mode
Tendency Not Top/ Very Strong Tendency Not Bottom	Adaptor Mode, but sometimes Stimulator Mode

Mover Mode

According to our theory, you often rely on Mover Mode if you scored over the average for both top-brain and bottom-brain processing. If you have a very strong tendency to use both top- and bottom-brain processing, you consistently operate in Mover Mode. If you have only a tendency (not a very strong tendency) to use both sorts of brain processing, then you may operate in Mover Mode but not as consistently. Instead, the mode you operate in may depend on the particular context in which you find yourself; if this mode is not strongly dominant, how you think and behave will depend in large part on the demands placed on you by your current situation.

If you have a very strong tendency to use top-brain processing but only a tendency to use bottom-brain processing, you typically are operating in Mover Mode with the possibility of sometimes slipping into Stimulator Mode. If you have a very strong tendency to use bottom-brain processing but only a tendency to use top-brain processing, you typically are operating in Mover Mode with the possibility of slipping into Perceiver Mode.

Perceiver Mode

You often operate in Perceiver Mode if you scored over the mean for bottom-brain processing but below the mean for top-brain processing. If you have a very strong tendency to use bottom-brain processing and a very strong tendency not to use top-brain processing, you are consistently in Perceiver Mode. If you have only a tendency to use bottom-brain processing and only a tendency not to use top-brain processing, then you may operate in Perceiver Mode but not as consistently—rather, your mode of thinking may depend on the context in which you find yourself.

If you have a very strong tendency to use bottom-brain processing but only a tendency not to use top-brain processing, you typically are operating in Perceiver Mode with the possibility of sometimes slipping into Mover Mode. If you have a very strong tendency not to use top-brain processing but only a tendency to use bottom-brain processing, you typically are operating in Perceiver Mode with the possibility of slipping into Adaptor Mode.

Stimulator Mode

According to our theory, you often rely on Stimulator Mode if you scored over the mean for top-brain processing but below the mean for bottom-brain processing. If you have a very strong tendency to use top-brain processing and a very strong tendency not to use bottom-brain processing, you are consistently in Stimulator Mode. If you have only a tendency to use top-brain processing and only a tendency not to use bottom-brain processing, then you may operate in Stimulator Mode but not as consistently—rather, the mode you rely on may depend on the context in which you find yourself.

If you have a very strong tendency to use top-brain processing but only a tendency not to use bottom-brain processing, you typically are operating in Stimulator Mode with the possibility of sometimes slipping into Mover Mode. If you have a very strong tendency not to use bottom-brain processing but only a tendency to use top-brain processing, you are operating in Stimulator Mode with the possibility of slipping into Adaptor Mode.

Adaptor Mode

You often operate in Adaptor Mode if you scored below the mean for both top-brain and bottom-brain processing. If you have a very strong tendency not to use both top- and bottom-brain processing, you are consistently in Adaptor Mode. If you have only a tendency not to use bottom-brain processing and only a tendency not to use top-brain processing, then you may operate in Adaptor Mode but not as consistently—rather, the mode you rely on may depend on the context in which you find yourself.

If you have a very strong tendency not to use top-brain processing but only a tendency not to use bottom-brain processing, you typically are operating in Adaptor Mode with the possibility of sometimes slipping into Perceiver Mode. If you have a very strong tendency not to use bottom-brain processing but only a tendency not to use top-brain processing, you typically are operating in Adaptor Mode with the possibility of slipping into Stimulator Mode.

Unlike many pop-psychology quizzes—the Spinning Dancer and Hemispheric Inventory tests, for example—this one was built on a solid foundation. Here's how it was developed.

The key was the characterization of what the top- and bottom-brain systems accomplish, as summarized earlier in this book. This led the developers to produce close to a hundred statements (of the same sort that eventually made it into the test), each of which they believed to characterize a preference, habit, or behavior that reflects mostly top-brain or mostly bottom-brain optional (not dictated by the immediate situation) processing. The survey service SurveyMonkey was used to give these items via the Internet to more than three hundred people (ranging in age from eighteen to sixty-five years old, with slightly more men than women, by the luck of the draw). These people indicated the degree to which they agreed or disagreed with each statement.

The scores were then correlated. Correlations indicate the degree to which one set of scores (for example, for a particular test item) varies in tandem with another (for example, scores for a different test item). As discussed earlier, the tighter the relation, the higher the correlation (within the range of 0 to 1.0 when the scores increase in tandem or 0 to –1.0 when one set of scores increases while the other decreases). If the two sets of scores are completely unrelated, the correlation will be 0.

The correlations for every pair of test items were obtained, and a mathematical technique known as factor analysis was applied, in the same way that this technique was used to develop the test of object versus spatial mental imagery we discussed in chapter 4. Factor analysis attempts to specify a set of dimensions, called factors, that give rise to a set of scores. That is, the pattern of correlations is taken to reflect the presence of relatively few underlying common influences; to the extent that the scores for two items are affected by the same influences, they will be positively correlated more strongly. For each item, the technique assigns a "loading" on the different factors, with higher numbers indicating that a given factor plays a stronger role in the correlations between the scores for that item and the scores for the other items.[2]

The crucial point is that we would expect two factors to underlie the scores for the statements: one that reflects the influence of the top-brain system and one that reflects the influence of the bottom-brain system. And that indeed was the case: Items that had high loadings on one factor and low loadings on the other were easily interpreted as drawing primarily on the top-brain system or primarily on the bottom-brain system.

However, there's an art to this science: The results of factor analysis are only as good as the items that are tested. The researchers spent six months testing and then revising new items. After each set of items was tested, it was usually obvious—in hindsight!—why certain items loaded on both underlying factors (reflecting both top-brain and bottom-brain processing), instead of loading on just a single one of those factors. Four iterations of test development were necessary over the course of six months to produce a set of items in which each item loaded primarily on only one of the underlying factors. At each iteration, new items had to be created and added to the list of the best previous ones (the new items often were refined versions of previous ones that were not satisfactory as originally written), new people tested, the results analyzed, and items deleted that didn't load primarily on a single factor.

After the factor analysis was complete, the correlations among the scores for each item were checked. The only items that were retained had scores that were more highly correlated with those of the other items in the same scale (top brain or bottom brain) than with the scores for the items in the opposite scale. This process eventually allowed acceptable items to be selected for the test.

It's important that for the final set of twenty items (included in the test), the average correlation between the top-brain items and the bottom-brain items was $r = .03$. This is a tiny, tiny correlation. It means that far less than one-hundredth of a percent of the variation in the scores for the top-brain items could be predicted by the variation in scores for the bottom-brain items, and vice versa. Thus, the two scales are measuring separate things.

Finally, another group was asked to take the test along with a variety of other, previously developed tests. These other tests included a test of general intelligence; a personality inventory,[3] which assesses the five dimensions that are consistently detected in studies of personality, namely openness (being open to new experience versus cautious), conscientiousness (being efficient and careful versus careless), extraversion (being outgoing versus reserved), agreeableness (being friendly and compassionate versus cold), and neuroticism (being nervous or sensitive versus confident and secure);[4] the Object-Spatial-Imagery-Verbal Questionnaire,[5] which assesses mental imagery and verbal abilities; and the Marlowe-Crowne Social Desirability Scale,[6] which assesses whether the test-taker tends to provide socially desirable responses.

The top-brain and bottom-brain scales had different patterns of correlation with the scores on other tests; this is further evidence that the scales of the new test do in fact tap different types of processing. Specifically, the scores on the bottom-brain scale did not correlate with any of the other test scores; this means that these scores are measuring something completely distinct. In contrast, the scores on the top-brain scale correlated with scores on an intelligence test—and that makes sense if intelligence involves strategic thinking. In addition, these scores correlated with some personality measures, such as "conscientiousness." These correlations also make sense, given that the top-brain scores reflect strategic thinking.

In short, all of this work paid off: The test that was finally developed and is presented here includes items that appear to tap different facets of top-brain and bottom-brain processing and that produce scores that can be easily interpreted, as you have just seen. However, we must note that although the test is reliable (it produces comparable scores when people are retested), it has not yet undergone rigorous validation—showing that scores do in fact predict what we would expect them to predict. At this juncture, the test and its scores must be approached with this understanding.

Nevertheless, we hope that you have learned something about yourself simply by taking this test and scoring your responses. It might be an interesting exercise to have others in your life do the same; you could compare your habitual cognitive modes and discuss them with one another. Who knows, you might gain insight into why your interactions sometimes work well and sometimes leave something to be desired.

So far we've presented a theory that is based on solid facts about the brain, and we've noted some of the implications of this theory. This could seem largely like an academic exercise. So why should you care about this particular theory, when there are so many other theories already out there? In the following section we draw out more implications of the Theory of Cognitive Modes—which we believe are sufficiently important to warrant taking the theory seriously. If our theory is correct, it will have a lot to say about important aspects of being human.

Chapter 14

Working with Others

As we have noted, you are not frozen in one mode of thinking at all times. Just as you may have a favorite beverage but still occasionally reach for another, so do you sometimes shift your mode of thought. But we also have discussed reasons why most of the time you probably will have a single dominant mode. According to the theory, our temperaments and our personal experiences both contribute to this tendency. Moreover, as we have seen, a good chunk of our temperaments is strongly influenced by our genes—and consequently is difficult to change. We also noted that you would need about ten thousand hours of practice to acquire enough knowledge in an area to be able to operate in Mover Mode effectively in that area—and even then, this expertise usually would apply only to that particular area. Becoming an expert coach at baseball won't help much in becoming an expert coach at hockey.

So what are we to do when we realize that our dominant mode is not appropriate for a particular situation? It often may not be easy simply to switch modes, no matter how much we would like to do so.

One strategy has proved to be an effective way to cope with

what we call *rigid mode syndrome*, which we define as occurring when one has great difficulty switching out of one's dominant cognitive mode and into another mode that is more appropriate for a specific situation. The following parable illustrates the strategy:

Some adult animals living in the forest were worried about the younger generation. The youngsters were hanging around the clearings and loitering on the trail corners and generally failing to develop their potential.

So the adults decided to start a school.

Their first job was designing a curriculum. The bears pointed out that digging was absolutely essential—it definitely needed to be on the "must-be-taught" list. The birds chirped in that flying could not be overlooked. The rabbits naturally emphasized running very fast, and so forth. In the end, all agreed that all of these skills were important, and thus every species should learn every one.

The adults gathered the young animals together and began their education. Before long, there were young birds with broken wingtips from trying to dig, baby bears with sprained ankles from trying to run very fast, and baby rabbits with bruises from trying to fly. Needless to say, once they recovered, none of the youngsters was happy or better-educated. The curriculum was a failure.

The moral of the story is *not* that some people are like birds, some like bears, and some like rabbits, and that each of us is best suited for certain tasks but not for others. Rather, the moral is that once you learn what kind of animal you are, you can more effectively approach a task. If you're a bird and want to dig, you use your beak and claws and realize that you would be very effective on an archeological site but less effective if you wanted to dig a den. If you're a bear, you should know that heavy digging is your thing—so if you want to dig, digging large holes is what you do best. And if you're a rabbit, you should know that running fast is what you do well—but if you want to fly, best to get on an airplane.

In other words, do what you can do well, and if you don't have an affinity to do what you need for a certain situation, seek someone who can collaborate with you.

Greebles in the Maze

The value of this strategy has been demonstrated in the lab.

In 2005, the so-called Group Brain Project began in earnest at Harvard. This group included many researchers but was led by the late J. Richard Hackman and Stephen.[1] By means of SurveyMonkey, the team screened over two thousand people online, using the Object and Spatial Imagery Questionnaire, described in chapter 4. The researchers invited two hundred of the people who responded to this questionnaire to come into the laboratory for further testing; these people had scored high in bottom-brain, object-based mental imagery (which allows one to visualize shapes and colors well) but low in top-brain, spatial-based mental imagery (which allows one to visualize locations in space)—or they had scored low in bottom-brain, object-based mental imagery but high in top-brain, spatial-based mental imagery.

Pairs of the selected people came to the lab at the same time and participated in a task that required navigating through a virtual maze that was shown on a computer screen. The maze was shown as if one were actually in it, and a joystick allowed the user to move forward down a corridor or turn to the left or right when reaching a branch point. At various locations in the maze, made-up objects called greebles sat, rooted to the floor. In some cases, the same greeble appeared a second time, later in the maze.[2]

The researchers defined two roles for each team and assigned one member of the team to each role. One person was asked to use the joystick to navigate the maze, and the other person was asked to use a button to tag duplicated greebles (that is, to indicate when a greeble encountered in the maze had exactly the same shape as one seen earlier). The teams were sent into a particular maze only once and had three minutes to navigate it and tag duplicate greebles; they were paid in proportion to how many greebles they correctly tagged.

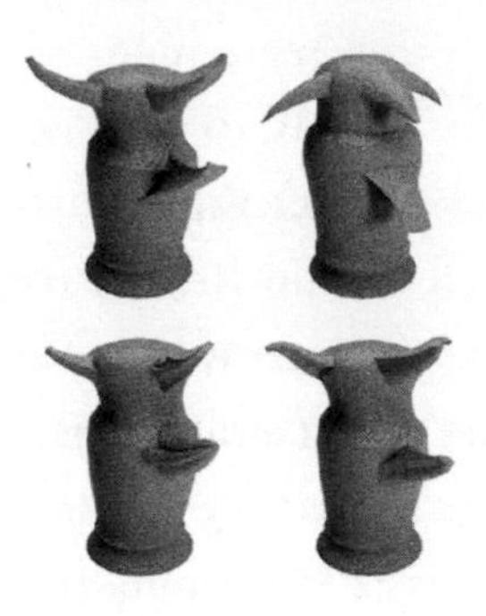

Computer-generated artificial objects, known as greebles, used as stimuli in the maze experiment. *Images courtesy of Michael J. Tarr, Carnegie Mellon University, www.tarrlab.org.*

Crucially, some pairs of participants included one member who had scored high on top-brain, spatial-based mental imagery but low on bottom-brain, object-based mental imagery. In these teams, the second member had the opposite sets of scores, high on object-based mental imagery but low on spatial-based mental imagery.[3]

Here is the trick of the experiment: The participants were either given roles that fit their strengths or given the opposite, incompatible, roles. That is, in the *compatible* condition the high-spatial-imagery person was asked to navigate and the high-object-imagery person (who is adept at classifying objects, as well as other sorts of stimuli) was asked to tag; in the *incompatible* condition, the role assignments were reversed. Finally, a third group contained either two high-object-imagery or two high-spatial-imagery people.

Here's what happened: The teams in which the assigned roles were compatible with the participants' abilities performed much better than the other two types of teams. *However, these were the findings when team members were not allowed to talk to each other.* When team members *were* allowed to talk to each other while navigating through a second maze, a different picture emerged:

First, the compatible teams performed comparably to how they performed when they were not allowed to talk, which is not surprising—each member functioned well on his or her own and didn't need any input from the other team member.

Second, when allowed to talk to each other during the task, the incompatible teams did far better than they did when they were not allowed to talk. Why? Videotapes of the sessions revealed that the high-spatial-imagery person soon took over navigation, telling the high-object-imagery (but low-spatial-imagery) team member where to turn, and the high-object-imagery person soon took over tagging, telling the low-object-imagery (but high-spatial-imagery) person when he or she had encountered a duplicate greeble.

The results were fascinating; without being told about their scores on the screening tests, and not having known each other before they came into the lab, the team members spontaneously discovered their relative strengths and weaknesses. And, when appropriate, they essentially switched roles to play to their strengths and avoid relying on their weaknesses.

But perhaps even more informative were the findings for the teams with either two high-object-imagery people or two high-spatial-imagery people: The more that the members of the team communicated with each other, the *worse* they did (that is, the fewer duplicate greebles were tagged during the allotted time). The blind were leading the blind. They lacked a key skill needed to get the job done but didn't realize it.

To return to the parable about the animals, it was as if a bear turned to another bear for guidance on how to fly.

Social Prosthetic Systems

Clearly, if you don't have the ability or skill to do something you need to do, you should turn to someone (or something) else for help. That much is obvious—or is it? How to explain why so many of us have trouble asking other people to help us? (The classic example, from the pre-GPS age, is how many men won't stop at a gas station to ask for directions.)

Our recommendation: First, overcome reluctance to ask for help. The question then becomes: Who, exactly, should you reach out to?

Answers can be found in the principles of what we call *social prosthetic systems*, a name coined by drawing an analogy to physical prosthetic systems.[4] Imagine that you had lost a leg. To help you walk, you would rely on a prosthesis—the modern-day steel-and-plastic equivalent of a wooden leg. This prosthesis makes up for a lack, allowing you to accomplish a task (in this case, walking). Not only do people rely on physical prostheses, but they also can rely on mental ones. If you are asked to multiply two large numbers (say, 7,481,222 × 1,532,596), you will want paper and pencil—or, better yet, a calculator. If you are a parent and have to keep track of your busy family's many activities, you will want a wall calendar or day book—or an event app for your phone. These devices serve as *cognitive prostheses*—they make up for a cognitive lack that must be filled in order for you to accomplish a specific task.

The Internet, of course, has evolved into what we might call the mother of all cognitive prostheses—the place many of us now turn, typically via Google and other search engines, to find facts, directions, images, translations, and more. We store personal data and cherished memories (in the form of photographs and videos) on the cloud, from which they can be easily (and precisely) retrieved. James Gleick, author of *The Information: A History, a Theory, a Flood,* calls the billions of Web pages that constitute the Internet "the global prosthetic brain."[5]

Despite its informational power, the Internet is of limited use when we need wise advice to help us navigate through a thorny situation. The main cognitive prosthesis we rely on for such help is not software or machines but other people: others who can help us extend our intelligence and discover and regulate our emotions. These are the people who constitute our social prosthetic systems. As Stephen defined it in the initial paper on the idea, social prosthetic systems are "human relationships that extend one's emotional or cognitive capacities. In such systems, other people serve as prosthetic devices, filling in for lacks in an individual's cognitive or emotional abilities."[6] And every person, with the possible exception of a committed hermit, belongs to one or more of these systems.

A crucial lesson from the animals-in-the-forest parable and the greebles-in-the-maze experiment is that you should select your social prostheses depending on the particular task you want to accomplish. By analogy, someone who is missing her feet can select from a host of different prostheses, which are more or less useful for different kinds of ambulation. Very springy metal feet can help a person run faster, but conventional plastic ones can be more appropriate when walking long distances. Similarly, depending on what you need to do cognitively or emotionally, and your own dominant cognitive mode, you may wish to be in the company of a person who is knowledgeable in the relevant area and tends to operate in a particular mode.

Let's say you were in an emotionally fraught situation—on the verge of breaking up with a spouse or partner. You would probably not want to seek the counsel of someone who typically operates in Stimulator or Adaptor mode. A person operating in Stimulator Mode might simply give a knee-jerk reaction, and a person operating in Adaptor Mode might try to minimize the issue. So that would leave you with the choice of counsel from someone who typically operates in Mover Mode or Perceiver Mode. And *that* choice would depend in part on your goals for the outcome. If you wanted strategic help in how to handle the situation, the theory suggests that the person in Mover Mode would be most appropriate. But if you wanted reflections on how you were actually feeling, and on what you wanted and needed, the person who typically operates in Perceiver Mode might be best. Putting this together, you might want to seek counsel from two separate people to garner the benefits of both kinds of input. Thus informed, you could more wisely make decisions.

Social prosthetic systems can be set up and maintained over many years. Over time, you may come to realize that certain people you know are ideal partners to help you in specific contexts. The political world gives us great examples, which nevertheless can be applied by the rest of us to everyday life—or business and work.

Take the mayor, governor, representative, or other official who has an efficient staff. His or her policy experts probably are people who habitually operate in Perceiver Mode; the person answering the constituent phone perhaps habitually uses Adaptor Mode; the chief of staff might be someone who often operates in Mover Mode or Stimulator Mode (if the latter, to be effective the chief of staff probably has assistants who operate in Perceiver Mode, to keep him or her from veering off course). All the while, the official could be operating in Mover Mode. She or he is at the center, drawing on help as needed—and preserving this rich system for the future as well as using it in the present.

We can also find this same sort of structure in highly functional families, with the matriarch or patriarch typically behaving in Mover Mode, a rambunctious child showing Stimulator Mode tendencies as she regales her relatives with tales of her adventures, and the kindly aunt or uncle who offers Perceiver Mode perspective to nieces and nephews seeking counsel.

Contemporary TV brings us another example. The creators of the Emmy Award–winning show *Modern Family* seem to have intuited the essentials of social prosthetic systems when they scripted the Dunphy family dynamic. Father Phil acts as if he operates in Stimulator Mode much of the time, always making plans that invariably fall apart when he does not properly react to changing circumstances—but he tries. Mother Claire often (though not always) acts as if she functions in Mover Mode, attempting to rein in her husband and children while keeping everyone on the same page. With her tendency to be drawn into one situation after another, older daughter Haley would seem frequently to behave in Adaptor Mode—but she adds value with her entertaining and playful spirit. Alex, Haley's younger sister, clearly acts as if she prefers to operate in Perceiver Mode, serving as the wiser-than-her-years commentator on her family's unfolding life—dispensing advice that, not surprisingly, is sometimes ignored. Although still young, son Luke seems to be headed toward a preference for operating in Stimulator Mode, like his father. Together, the Dunphys make it work. By the end of each episode, they have resolved issues and moved ahead, the family still functioning as a unit.

The crucial idea is that when you are interacting with another person in this way, he or she has the capacity to make you more effective, in the role of a "social prosthetic." He or she fills in for your lack. And in the process, at that moment you become a different person, transformed by your interactions with the other, just as the amputee who wears the springy artificial feet becomes a better runner than when she wears the conventional artificial foot.

Humor us by considering one more analogy. When you put a chopping blade into a Cuisinart it becomes a different machine from what it is when you put a blending blade into it—with the first being good for dicing apples for a pie, the second for making smoothies. When you are relying on another person to function as your social prosthesis, the two of you together function as something different from what either of you is alone or with companions who have different skills and abilities. It's not as if you are merely seeking a consultant or an aide: If you interact closely with someone who knows you well, he or she can fill in what you are missing. In this case, the whole becomes more than the sum of its parts. It's still you who are setting the agenda and driving the train (we are not talking about a conventional team), but you are now augmented by a companion's knowledge and skill set—which allows you to do more than you could on your own. It's as if you have borrowed part of your companion's brain, thereby extending your own reach and capacities.

Thinking Twice

Ideally, you would have time to reflect on the perspectives presented in this book before beginning a new job or entering a new social or personal relationship. According to the theory (and we again remind you that it is a theory), someone who was prone to being in Stimulator Mode might want to pause to think carefully about marrying someone who also was prone to being in Stimulator Mode—such a union could easily produce a marriage rife with conflict. Two people who typically operate in Adaptor Mode might want to reflect on what a marriage would be like if much of the time "nothing is happening" or the situation is constantly careening or being buffeted by events. Imagine how things might be different for Nick and Erica, the characters we created for chapter 12, if Erica, like her husband, habitually thought and behaved in Adaptor Mode and not in her dominant Mover Mode; with the demands of running the household and managing the couple's three young children, it might be chaotic.

Although two people who habitually operate in Perceiver Mode might have a low-stress relationship, achieving goals that require detailed or complex planning could prove challenging. Picture Hannah and Rick, the characters from chapter 10. If Rick habitually thought and behaved in Perceiver Mode, as Hannah does, the two of them would enjoy a comfortable life—but as they neared their sixties, would either of them have made the financial decisions necessary to ensure a comfortable retirement? The point is not that Hannah (or any librarian) is incapable of long-term financial decision making, only that this is not likely to be her natural inclination—but it is where Rick's dominant cognitive mode advances the couple's shared interests.

Similarly, the theory leads us to expect that if a person habitually operates in Perceiver Mode in the professional world, it might be most comfortable for him or her to work with people who often are in Perceiver or Adaptor modes. But, comfort aside, this often would probably be less than ideal. Arguably, most teams would benefit by having some members who are comfortable and adept in Mover Mode, others who are comfortable and adept in Adaptor Mode, and so on. For example, people who prefer to operate in Perceiver Mode would get a lot out of working with those who prefer Mover or Stimulator modes, and vice versa.

In order to change your dominant mode, you need to be highly motivated, have a lot of time, and stick to the effort—and even then, this change will probably affect your functioning only in a particular domain. Not everybody is so patient. In most cases, we suspect that you probably would be better off identifying your dominant cognitive mode and finding people who have dominant modes that complement your own. And remember that a person's mode may be different in different circumstances (which draw on different sets of knowledge)—a person comfortable with Mover Mode at work may be most comfortable in Adaptor Mode at home, and a person who usually operates in Stimulator Mode with friends may slide into Perceiver Mode with a mate. Thus, if our theory is on the right track, be sure to spend time with a person in the appropriate circumstances if you are seeking compatibility.

• • •

Some readers of this book will find themselves already in problematic situations. What then? The Theory of Cognitive Modes cannot provide definitive guidance, but knowing about the four modes can make you sensitive to certain potential problems before you become involved with someone. Moreover, the theory implies that you can become an expert on someone close to you. And learning to predict his or her likely reactions can help you operate in Perceiver and Mover modes, which can make a difficult situation manageable. If you don't have the motivation or time to learn how to cope, seek a friend (or counselor) who can complement your strengths, filling in for what you cannot do easily. Here, again, would be the value of relying on an appropriate social prosthetic system.

Working well with others is arguably the most important thing most of us do. There are two clear keys to success: The first is to grow, by learning new strategies (ways to plan and behave, using the top-brain system) and learning new ways to "frame" a situation (ways to classify and interpret, using the bottom-brain system). The second is to change your circumstances, whether work, home, or social setting. In any given situation, you can use one of these two keys to open a new door.

Being One's Own Self

The great French Renaissance thinker Michel de Montaigne observed: "The greatest thing in the world is to know how to be one's own self." In this echo of Lao Tzu's ancient observation, with which we opened this book, Montaigne implores us each to look within, into our own unique character. Our character arises fundamentally from our brains, steeped in experience. And each of us should look within not just fleetingly, we would argue, but with a committed and lifelong purpose that can be its own reward.

We hope that *Top Brain, Bottom Brain* will be helpful to you in your own journey of discovery. We hope that the ideas presented in these pages will stimulate you to find new perspectives about yourself and the people you meet on your way.

"The value of life lies not in the length of days, but in the use we make of them," Montaigne wrote.

Authors' Note

This collaboration began one autumn afternoon in University Hall at Harvard University, where Stephen was dean of social science. Among his responsibilities was running his laboratory in the psychology department and being associate psychologist in the Department of Neurology at the Massachusetts General Hospital. Wayne was no stranger to Harvard, having graduated from the college and written two earlier books about pioneering researchers at the Harvard Medical School and Boston Children's Hospital, a Harvard affiliate.

Stephen was looking for a writing partner to help bring his new Theory of Cognitive Modes to a wide readership. Wayne was intrigued by the theory and the underlying science, and by the contention that the popular left brain–right brain story was just that, a story; like many people, he had assumed it was natural law. But he was hardly convinced that Stephen, a clear writer as well as a scientist, needed anyone's assistance. Stephen begged to differ; he felt that he could write clearly but had no idea how to write for a general audience. That initial meeting led to a longer session another day in Stephen's other office, on the eighth floor of William James Hall—a session that was followed by dinner. There, scientist and writer agreed this would work.

It *did* work—and it *was* work, a lot of it, spanning many months and eventually a continent, when Stephen left Harvard for Stanford (and, then, more recently, for the new university being born in San Francisco, as part of

Minerva Project), leaving Wayne in New England. The authors met several times, in New York and elsewhere, but the book evolved primarily around hundreds of emails and telephone conversations—and, of course, the solitary writing. Back and forth went ideas, which led to outlines, which became drafts, which became more drafts, which through editing became the book you are reading. There were spirited debates over word choice and sentence selection, and weightier discussions of substance and nuance, and what was "too much" science to include and what was "too little"—that is, there were questions of balance. All of this suggests that we both frequently were operating in Mover Mode—with occasional forays into Stimulator thinking. And truth be told, there were also times when Stephen transitioned into professorial Perceiver Mode, and Wayne fell into Adaptor Mode (in dutifully editing Stephen's prose to make it accessible and interesting).

Given that Stephen comes from the scientific world and Wayne from the public square, inevitably there were disagreements—but invariably, they were gentlemanly, and always resolved amicably. Neither of us can recall a single argument, or even a heated conversation (or email), though perhaps memory fails us. We had a common purpose and a clear sense of each other's strengths. And although we never really discussed our collaboration as such, we really became a living example of social prosthetics: with Wayne providing the right turn of phrase when Stephen came up short, and Stephen providing the scientific expertise that Wayne lacked.

More than that, a friendship developed . . . and with it, the hope of further collaborations.

The authors were brought together by agent Rafe Sagalyn, who represented Stephen; and Jon Karp, now Simon & Schuster publisher, who edited four of Wayne's books, including two nonfiction narratives set in the world of medicine, while Jon was at Random House. So our joint gratitude to Jon and Rafe, for their faith in us. Gratitude, too, to Shannon O'Neill and Rebecca Sagalyn at ICM®/Sagalyn. Thanks to editor Karen Thompson Walker, who left Simon & Schuster after acquiring the book to devote her full energies to her remarkable debut novel, *The Age of Miracles.* Editor Karyn Marcus inherited our manuscript, and from there, she graciously and patiently guided us. Thank you, Karyn—your suggestions on structure and style were invariably on target. At Simon & Schuster, thanks also to assistant editor Nicholas Greene, former associate marketing manager Rachelle Andujar and her successors Stephen Bedford and Marie Kent, publicist Kate Gales, jacket designer Christopher Lin, interior designer

Ruth Lee-Mui, copy editor Janet Byrne, indexer Judith Hancock, and counsel Emily Remes. This has been one great crew to have as partners!

We gratefully thank friends and colleagues who took valuable time to read an earlier version of this book or discuss the key ideas. Rusty Bobrow, Jon Cox, Jeffrey Epstein, Dan Gilbert, David Kosslyn, Justin Kosslyn, Steven Pinker, and Robin Rosenberg helped us enormously. Our thanking these people does not necessarily mean that they agree with the final product!

Individually, Stephen would like to thank his wife and collaborator, Robin Rosenberg (clinical psychologist extraordinaire, textbook author, and the go-to psychologist for the psychology of superheroes), and his collaborators who made this work possible, in particular William Thompson and Grégoire Borst.

Wayne extends his gratitude first and foremost to Yolanda Gabrielle, for her support and tolerance of his writing obsession—and her own insights into cognitive behavior, which spring from her profession as a therapist. At Salve Regina University he thanks Sisters Jane Gerety and M. Therese Antone, and also Kristine Hendrickson, and Jim Ludes and Teresa Haas of the Pell Center for International Relations and Public Policy. He also thanks those at *The Providence Journal* who have supported his writing for so long: Howard Sutton, Tom Heslin, Karen Bordeleau, Sue Areson, John Kostrzewa, Tom Mooney, Bob Kerr, and Bill Reynolds. Also, my screenwriting partner, Drew Smith. And finally, thanks to Michael Prevett, his long-time Los Angeles agent and friend.

—Stephen M. Kosslyn and G. Wayne Miller

Notes

Chapter 1: A New Way of Looking At What Your Brain Says About You

1. R. F. Thompson, *The Brain: A Neuroscience Primer*, 2nd ed. (New York: W. H. Freeman, 1993).
2. Adapted from F. A. Wilson, S. P. Scalaidhe, and P. S. Goldman-Rakic,"Dissociation of Object and Spatial Processing Domains in Primate Pre-Frontal Cortex," *Science* 260 (June 1993): 1955–58.

Chapter 2: Roots of the Theory

1. See Swedenborg's two-volume *The Economy of the Animal Kingdom*, 1740–1741 (1918).
2. W. Penfield and H. Jasper, *Epilepsy and the Functional Anatomy of the Human Brain* (Boston: Little, Brown, 1951).

Chapter 3: The Duplex Brain

1. Because vision and audition are the primary senses for us humans, we focus on them here; however, we note that comparable structures exist for the other senses.
2. For a good overview of the functions of this area, see: http://en.wikipedia.org/wiki/Broca's_area.
3. However, this separation is not absolute; under some circumstances, shape can be specified by location (think, for example, of how you

can "see" a triangle when three dots, specifying the locations of the vertices, are present) and locations of specific parts of shapes can be attended to; see S.R. Lehky, X. Peng, C. J. McAdams, and A. B. Sereno,"Spatial Modulation of Primate Inferotemporal Responses by Eye Position," *PLoS ONE 3(10)*: e3492. doi:10.1371/journal.pone.0003492; A. B. Sereno and S. C. Amador, "Attention and Memory-Related Responses of Neurons in the Lateral Intraparietal Area During Spatial and Shape-Delayed Match-to-Sample Tasks," *Journal of Neurophysiology* 95 (2006): 1078–98.

4. A. Treisman and H. Schmidt, "Illusory Conjunctions in the Perception of Objects," *Cognitive Psychology* 14 (1982): 107–41.
5. J. G. Rueckl, K. R. Cave, and S. M. Kosslyn, "Why Are 'What' and 'Where' Processed by Separate Cortical Visual Systems? A Computational Investigation," *Journal of Cognitive Neuroscience* 1 (1989): 171–86.
6. Melvyn Goodale and A. David Milner "Separate Visual Pathways for Perception and Action," *Trends in Neurosciences* 15 (1992): 20–25.
7. S. M. Kosslyn, "You Can Play 20 Questions with Nature and Win: Categorical Versus Coordinate Spatial Relations as a Case Study," *Neuropsychologia* 44 (2006): 1519–23.
8. G. Borst, W. L. Thompson, and S. M. Kosslyn, "Understanding the Dorsal and Ventral Systems of the Human Cerebral Cortex: Beyond Dichotomies," *American Psychologist* 66, no. 7 (2011): 624–32.

Chapter 4: Reasoning Systems

1. S. M. Kosslyn, W. L. Thompson, and G. Ganis, *The Case for Mental Imagery* (New York: Oxford University Press, 2006).
2. D. N. Levine, J. Warach, and M. J. Farah, "Two Visual Systems in Mental Imagery: Dissociation of 'What' and 'Where' in Imagery Disorders Due to Bilateral Posterior Cerebral Lesions," *Neurology* 35 (1985): 1010–18.
3. As of early 2013.
4. M. Kozhevnikov, M. Hegarty, and R. E. Mayer, "Revising the Visualizer-Verbalizer Dimension: Evidence for Two Types of Visualizers," *Cognition and Instruction* 20 (2002): 47–77.
5. M. Kozhevnikov, S. M. Kosslyn, and J. Shephard, "Spatial Versus Object Visualizers: A New Characterization of Cognitive Style," *Memory and Cognition* 33 (2005): 710–26.
6. Ibid., 721.
7. O. Blazhenkova, M. Kozhevnikov, and M. A. Motes, "Object-Spatial

Imagery: A New Self-Report Imagery Questionnaire," *Applied Cognitive Psychology* 20 (2006): 239–63.

8. Ibid., 243.
9. O. Blazhenkova and M. Kozhevnikov, "Visual-Object Ability: A New Dimension of Nonverbal Intelligence," *Cognition* 117 (2010): 276–301.
10. Ibid., 283.
11. D. Kahneman, *Thinking, Fast and Slow* (New York: Farrar, Straus & Giroux, 2011).
12. A. W. Woolley, J. R. Hackman, T. E. Jerde, C. F. Chabris, S. L. Bennett, and S. M. Kosslyn, "Using Brain-Based Measures to Compose Teams: How Individual Capabilities and Team Collaboration Strategies Jointly Shape Performance," *Social Neuroscience* 2 (2007): 96–105.

Chapter 5: Sweeping Claims

1. R. W. Sperry, "Cerebral Organization and Behavior: The Split Brain Behaves in Many Respects Like Two Separate Brains, Providing New Research Possibilities," *Science* 133 (1961): 1749–57.
2. It's not left eye versus right eye but rather the *side* of each eye that's critical.
3. *Life* Magazine, October 1, 1971, p. 43. The extract below this quote is also from p. 43.
4. Sperry, who died in 1994, was mentor to Antonio E. Puente, professor of psychology, University of North Carolina at Wilmington, and also Sperry's biographer. We asked him whether Sperry intended his research to give birth to the pop-culture left/right story. He wrote: "The most 'outrageous' comment Roger Sperry ever made that I heard was made at the acceptance of the American Psychological Association Lifetime Achievement Award. In that brief speech he said something along the lines that his left hemisphere could not adequately express the emotions that his right hemisphere [was] feeling. We have the www.rogersperry.info website, which gathers a great deal of traffic. That website produces queries along the lines that you stated. And, honestly, they are so far from the mark that it is difficult to understand where they come from. His writings were as conservative as his science. What was not was his vision and that he rarely shared."
5. R. W. Sperry, "Consciousness, Personal Identity and the Divided Brain," *Neuropsychologia* 22 (1984): 661–73.
6. See http://www.youtube.com/watch?v=i-yhtXAzYwc. There are many other postings. The spinning dancer was revealed to be an optical

illusion by Steven Novella, Yale University School of Medicine clinical neurologist, on October 11, 2007, in his article "Left Brain–Right Brain and the Spinning Girl," on his NeuroLogicaBlog: Your Daily Fix of Neuroscience, Skepticism and Critical Thinking. The correct description of the spinning dancer as optical illusion was more widely publicized in the *New York Times* on April 28, 2008, in a blog posting by Tara Parker-Pope, who wrote: "While the dancer does indeed reflect the brain savvy of its creator, Japanese Web designer Nobuyuki Kayahara, it is not a brain test. Instead, it is simply an optical illusion called a reversible, or ambiguous, image."

7. See www.wherecreativitygoestoschool.com/vancouver/left_right/rb_test.htm.
8. See http://homeworktips.about.com/library/brainquiz/bl_leftright brain_quiz.htm.
9. See www.squidoo.com/braintest.
10. See www.amazon.com/Brainy-Baby-Inspires-Logical-Thinking/dp/B000063UYK/ref=sr_1_4?s=movies-tv&ie=UTF8&qid=1315481288&sr=1-4 and www.amazon.com/Brainy-Baby-Right-Brain/dp/B000063UYL/ref=pd_cp_mov_1.
11. See www.nurtureminds.com/soroban-abacus.htm.
12. See www.mastermindabacus.com/blog_abacus/abacus-stimulates-whole-brain-development/.
13. See http://www.youngevity.net/product/RG037_Brain.html.
14. About a third of left-handers have the reverse organization, with the left hemisphere performing the functions carried out by the right hemisphere of right-handed people, and vice versa; for simplicity of exposition, we will not reacknowledge this qualification throughout the text.
15. Winfrey described herself as "a right-brain kind of person" in the December 2008 issue of *O: The Oprah Magazine.*
16. Daniel H. Pink, *A Whole New Mind: Moving from the Information Age to the Conceptual Age. Why Right-Brainers Will Rule the Future* (New York: Riverhead, 2005).
17. Brenda Milner, "Interhemispheric Differences in the Localization of Psychological Processes in Man," *British Medical Bulletin* 27 (1971): 272–77.
18. S. Harnad and H. Steklis, "Comment on J. Paredes and M. Hepburn's 'The Split Brain and the Culture-and-Cognition Paradox,'" *Current Anthropology* 17 (1976): 320–22.

Chapter 6: Interacting Systems

1. H. Damasio, T. Grabowski, R. Frank, A. M. Galaburda, and A. R. Damasio, "The Return of Phineas Gage: Clues About the Brain from the Skull of a Famous Patient," *Science* 264 (1994): 1102–5, doi:10.1126/science.8178168].
2. R. L. Gregory, "The Brain as an Engineering Problem," in *Current Problems in Animal Behaviour*, edited by W. H. Thorpe and O. L. Zangwill (Cambridge: Cambridge University Press, 1961).

Chapter 8: Origins of the Modes: Nature Versus Nurture

1. For reviews, see R. Plomin et al., *Behavioral Genetics*, 6th ed. (New York: Worth Publishing, 2012); K. J. Saudino, "Behavioral Genetics and Child Temperament," *Journal of Developmental and Behavioral Pediatrics* 26 (2005): 214–23.
2. J. Kagan and N. Snidman, "Early Childhood Predictors of Adult Anxiety Disorders," *Biological Psychiatry* 46 (1999): 1536–41; J. Kagan, *Galen's Prophecy: Temperament in Human Nature* (New York: Basic Books, 1994).
3. M. H. McManis, J. Kagan, N. C. Snidman, and S. A. Woodward, "EEG Asymmetry, Power, and Temperament in Children," *Developmental Psychobiology* 41 (2002): 169–77.
4. T. J. Bouchard, "Genetic Influence on Human Psychological Traits: A Survey," *Current Directions in Psychological Science* 13 (2004): 148–51.
5. Ruth's lifetime batting average was .342; Williams's, .344.
6. I. Biederman and M. M. Shiffrar, "Sexing Day-Old Chicks: A Case Study and Expert Systems Analysis of a Difficult Perceptual-Learning Task," *Journal of Experimental Psychology: Learning, Memory, and Cognition* 13 (1987): 640–45.
7. K. A. Ericsson, W. G. Chase, and S. Faloon, "Acquisition of a Memory Skill," *Science* 208 (1980): 1181–82.

Chapter 11: Stimulator Mode

1. A summary of several polls on Sarah Palin and the 2008 presidential election is found at www.pollingreport.com/wh08.htm.
2. Palin's comments were widely reported. Among other places, a full text was published at: http://thepage.time.com/2011/01/12/palin-journalists-and-pundits-tried-to-manufacture-a-blood-libel/.

Chapter 12: Adaptor Mode

1 J. M. Weiss, "Effects of Coping Behavior in Different Warning Signal Conditions on Stress Pathology in Rats," *Journal of Comparative and Physiological Psychology* 77 (1971): 1–13; J. M. Weiss, "Effects of Coping Behavior With and Without a Feedback Signal on Stress Pathology in Rats," *Journal of Comparative and Physiological Psychology* 77 (1971): 22–30.

Chapter 13: Test Yourself

1. S. M. Kosslyn and W. L. Thompson, "Assessing Habitual Use of Dorsal Versus Ventral Brain Processes," *Biologically Inspired Cognitive Architectures* 2 (2012; e-pub August 9, 2012): 68–76, doi.org/10.1016/j.bica.2012.07.007.
2. For further information about factor analysis, see: http://mplab.ucsd.edu/tutorials/FactorAnalysis.pdf; http://rtutorialseries.blogspot.com/2011/10/r-tutorial-series-exploratory-factor.html; and http://www.chem.duke.edu/~clochmul/tutor1/factucmp.html.
3. S. D. Gosling, P. J. Rentfrow, and W. B. Swann, "A Very Brief Measure of the Big-Five Personality Domains," *Journal of Research in Personality* 37 (2003): 504–28, doi: 10.1016/S0092-6566(03)00046-1.
4. L. R. Goldberg, "An Alternative 'Description of Personality': The Big-Five Factor Structure," *Journal of Personality and Social Psychology* 59 (1990): 1216–29, doi: 10.1037/0022-3514.59.6.1216; R. R. McCrae and O. P. John, "An Introduction to the Five-Factor Model and Its Applications," *Journal of Personality* 60 (1992): 175–215, doi: 10.1111/j.1467-6494.1992.tb00970.
5. O. Blazhenkova and M. Kozhevnikov, "The New Object-Spatial-Verbal Cognitive Style Model: Theory and Measurement," *Applied Cognitive Psychology* 23 (2009): 638–63.
6. D. P. Crowne and D. Marlowe, "A New Scale of Social Desirability Independent of Psychopathology," *Journal of Consulting Psychology* 24 (1960): 349–54.

Chapter 14: Working with Others

1. A. W. Woolley, J. R. Hackman, T. E. Jerde, C. F. Chabris, S. L. Bennett, and S. M. Kosslyn, "Using Brain-Based Measures to Compose Teams: How Individual Capabilities and Team Collaboration Strategies Jointly Shape Performance," *Social Neuroscience* 2 (2007): 96–105.

2. These computer-generated shapes were created by Scott Yu, then at Yale University, who was supervised by Michael J. Tarr and Isabel Gauthier, now at Carnegie Mellon University and Vanderbilt University, respectively. For more information, see http://www.psy.vanderbilt.edu/faculty/gauthier/FoG/Greebles.
3. The new cognitive modes test, presented in the previous chapter, did not exist at the time of this experiment. It characterizes the top- and bottom-brain systems as a whole, not just by the type of mental imagery that arises from each system. Hence, we cannot say with certainty that these people in general relied on Stimulator Mode and Perceiver Mode.
4. See S. M. Kosslyn, "On the Evolution of Human Motivation: The Role of Social Prosthetic Systems," in *Evolutionary Cognitive Neuroscience*, edited by S. M. Platek, T. K. Shackelford, and J. P. Keenan (Cambridge, MA: MIT Press, 2006), 541–54; and S. M. Kosslyn, "Social Prosthetic Systems and Human Motivation: One Reason Why Cooperation Is Fundamentally Human," in *Evolution, Games and God: The Principle of Cooperation*, edited by S. Coakley and M. Nowak (Cambridge, MA: Harvard University Press, 2013).
5. *New York Times*, August 5, 2012.
6. S. M. Kosslyn, "On the Evolution of Human Motivation: The Role of Social Prosthetic Systems."

Bibliography

Chapter 1: A New Way of Looking at What Your Brain Says About You

Kosslyn, Stephen M., and Robin S. Rosenberg. *Introducing Psychology: Brain, Person, Group.* Boston: Pearson Learning Solutions, 2011.

Thompson, Richard F. *The Brain: A Neuroscience Primer*, 2nd ed. New York: W. H. Freeman, 1993.

Ungerleider, Leslie G., and Mortimer Mishkin. "Two Cortical Visual Systems." In David J. Ingle, Melvyn A. Goodale, and Richard J. W. Mansfield, eds., *Analysis of Visual Behavior*, 549–86. Cambridge, MA: MIT Press, 1982.

Wilson, Fraser, Séamas Scalaidhe, and Patricia Goldman-Rakic. "Dissociation of Object and Spatial Processing Domains in Primate Pre-Frontal Cortex." *Science* 260 (June 1993): 1955–58.

Chapter 2: Roots of the Theory

Clarke, Edwin, Kenneth Dewhurst, and Michael Jeffrey Aminoff. *An Illustrated History of Brain Function: Imaging the Brain from Antiquity to the Present.* San Francisco: Norman Publishing, 1996.

Dronkers, Nina F., Odile Plaisant, Marie-Thérèse Iba-Zizen, and Emmanuel A. Cabanis. "Paul Broca's Historic Cases: High Resolution MR Imaging of the Brains of Leborgne and Lelong." *Brain: A Journal of Neurology* 5 (2007): 1432–41.

Finger, Stanley. *Origins of Neuroscience: A History of Explorations into Brain Function*. New York: Oxford University Press, 1994.

Gall, F. J., and J. G. Spurzheim, *Anatomie et physionomie du système nerveux en général et du cerveau en particulier. Premier volume*. Paris, F. Schoell, 1810; F. J. Gall and J. G. Spurzheim, vol. 2, 1812; F. J. Gall, vol. 3, 1818; F. J. Gall, vol. 4, 1819.

Gross, Charles G. "Early History of Neuroscience." In *Encyclopedia of Neuroscience*, edited by George Adelman. Cambridge, MA: Birkhäuser Boston, 1987.

———. "Aristotle on the Brain." *The Neuroscientist* 1 (1995): 245–50.

Hebb, Donald O. *The Organization of Behavior: A Neuropsychological Theory*. New York: Wiley, 1949.

James, William. *The Principles of Psychology*. New York: Henry Holt, 1890. Available online at, among other places, http://archive.org/details/theprinciplesofp01jameuoft.

Kosslyn, Stephen M. "Seeing and Imagining in the Cerebral Hemispheres: A Computational Approach." *Psychological Review* 94 (1985): 148–75.

——— "You Can Play 20 Questions with Nature and Win: Categorical Versus Coordinate Spatial Relations as a Case Study." *Neuropsychologia* 44 (2006): 1519–23.

Laeng, Bruno. (1994). "Lateralization of Categorical and Coordinate Spatial Functions: A Study of Unilateral Stroke Patients." *Journal of Cognitive Neuroscience* 6 (1994): 189–203.

Molnar, Zoltán. "Thomas Willis (1621–1675), the Founder of Clinical Neuroscience." *Nature Reviews Neuroscience* 5 (2004): 329–35.

Morse, Minna. "The Much-Maligned Theory of Phrenology Gets a Tip of the Hat from Modern Neuroscience." *Smithsonian*, October 1997. www.smithsonianmag.com/history-archaeology/object_oct97.html.

O'Connor, James P. B. "Thomas Willis and the Background to *Cerebri Anatome*." *Journal of the Royal Society of Medicine* 96 (2003): 139–53.

Pevsner, Jonathan. "Leonardo da Vinci's Contributions to Neuroscience." *Trends in Neuroscience* 25 (2002): 217–20.

Pohl, Walter. "Dissociation of Spatial Discrimination Deficits Following Frontal and Parietal Lesions in Monkeys." *Journal of Comparative and Physiological Psychology* 82 (1973): 227–39.

Rahimi, Scott Y., Dennis E. McDonnell, Amir Ahmadian, and John R. Vender. "Medieval Neurosurgery: Contributions from the Middle East, Spain and Persia." *Neurosurgical Focus* 23 (2007): 1–4.

Rueckl, Jay G., Kyle R. Cave, and Stephen M. Kosslyn. "Why Are 'What' and 'Where' Processed by Separate Cortical Visual Systems? A Computational Investigation." *Journal of Cognitive Neuroscience* 1 (1989): 171–86.

Treisman, Anne, and Garry Gelade. "A Feature-Integration Theory of Attention." *Cognitive Psychology* 12 (1980): 97–136.

These works help inform the historical perspective of this book:

Allen, Richard. "David Hartley." *The Stanford Encyclopedia of Philosophy*, edited by Edward N. Zalta. 2009, http://plato.stanford.edu/archives/sum2009/entries/hartley/.

Arnott, Robert, Stanley Finger, and Christopher Upham Murray Smith, eds. *Trepanation: History, Discovery, Theory.* Lisse, Netherlands: Swets & Zeitlinger, 2005.

Broca skull fragment photo: http://scienceblogs.com/neurophilosophy/2008/01/an_illustrated_history_of_trep.php.

Hooke, Robert. *Micrographia: Some Physiological Descriptions of Minute Bodies Made by Magnifying Glasses with Observations and Inquiries Thereupon.* Project Gutenberg eBook #15491, released 2005. Originally published in 1665.

Photographs of the Smith Papyrus: www.neurosurgery.org/cybermuseum/pre20th/epapyrus.html.

Rahimi, Scott Y., Dennis E. McDonnell, Amir Ahmadian, and John R. Vender. "Medieval Neurosurgery: Contributions from the Middle East, Spain, and Persia." *Neurosurgical Focus* 23 (2007): 1–4.

Wade, Nicholas, Marco Piccolino, and Adrian Simmons. "Alessandro Volta: 1745–1827." In *Portraits of European Neuroscientists.* http://neuroportraits.eu/. 2011.

Wilkins, Robert H. "Neurosurgical Classic-XVII: Edwin Smith Surgical Papyrus." *Journal of Neurosurgery* 114 (1964): 240–44.

Chapter 3: The Duplex Brain

Borst, Grégoire, William L. Thompson, and Stephen M. Kosslyn. "Understanding the Dorsal and Ventral Systems of the Human Cerebral Cortex: Beyond Dichotomies." *American Psychologist* 66, no. 7 (2011): 624–32.

Collice, Massimo, Rosa Collice, and Alessandro Riva. "Who Discovered the Sylvian Fissure?" *Neurosurgery* 63 (2008): 623–28.

Goodale, Melvyn A., and A. David Milner. "Separate Visual Pathways for Perception and Action." *Trends in Neurosciences* 15 (1992): 20–25.

Lehky, Sidney R., Xinmiao Peng, Carrie J. McAdams, and Anne B. Sereno. "Spatial Modulation of Primate Inferotemporal Responses by Eye Position." *PLoS ONE 3(10)*: e3492. doi:10.1371/journal.pone.0003492.

MacLean, Paul D. *The Triune Brain in Evolution: Role in Paleocerebral Functions*. New York: Plenum Press, 1990.

Sereno, Anne B., and Silvia C. Amador. "Attention and Memory-Related Responses of Neurons in the Lateral Intraparietal Area During Spatial and Shape-Delayed Match-to-Sample Tasks." *Journal of Neurophysiology*, 95 (2006): 1078–98.

Sylvius, Franciscus. *Disputationem Medicarum*. 1663.

Chapter 4: Reasoning Systems

Aginsky, Vlada, Catherine Harris, Ronald A. Rensink, and Jack Beusmans. "Two Strategies for Learning a Route in a Driving Simulator." *Journal of Environmental Psychology* 17 (1997): 317–31.

Blazhenkova, Olesya, Maria Kozhevnikov, and Michael A. Motes. "Object-Spatial Imagery: A New Self-Report Imagery Questionnaire." *Applied Cognitive Psychology* 20 (2006): 239–63.

Blazhenkova, Olesya, and Maria Kozhevnikov. "The New Object-Spatial-Verbal Cognitive Style Model: Theory and Measurement." *Applied Cognitive Psychology* 23 (2009): 638–63.

———. "Visual-Object Ability: A New Dimension of Nonverbal Intelligence." *Cognition* 117 (2010): 276–301.

Blazhenkova, Olesya, Maria Kozhevnikov, and Michael Becker. "Object-Spatial Imagery and Verbal Cognitive Styles in Children and Adolescents." *Learning and Individual Differences* 21 (2011): 281–87.

Ekstrom, Ruth B., John W. French, and Harry H. Harman. *Manual for Kit of Factor Referenced Cognitive Tests*. Princeton, NJ: Educational Testing Service, 1976.

Kahneman, Daniel. *Thinking, Fast and Slow*. New York: Farrar, Straus & Giroux, 2011.

Kosslyn, Stephen M., William L. Thompson, and Giorgio Ganis. *The Case for Mental Imagery*. New York: Oxford University Press, 2006.

Kozhevnikov, Maria, Mary Hegarty, and Richard E. Mayer. "Revising the Visualizer-Verbalizer Dimension: Evidence for Two Types of Visualizers." *Cognition and Instruction* 20 (2002): 47–77.

Kozhevnikov, Maria, Stephen M. Kosslyn, and Jennifer Shephard. "Spatial Versus Object Visualizers: A New Characterization of Cognitive Style." *Memory and Cognition* 33 (2005): 710–26.

Levine, David N., Joshua Warach, and Martha J. Farah. "Two Visual Systems in Mental Imagery: Dissociation of 'What' and 'Where' in Imagery Disorders Due to Bilateral Posterior Cerebral Lesions." *Neurology* 35 (1985): 1010–18.

Motes, Michael A., Rafael Malach, and Maria Kozhevnikov. "Object-Processing Neural Efficiency Differentiates Object from Spatial Visualizers." *NeuroReport* 19 (2008): 1727–31.

Rouw, Romke, Stephen M. Kosslyn, and Ronald Hamel. "Detecting High-Level and Low-Level Properties in Visual Images and Visual Percepts." *Cognition* 63 (1997): 209–26.

Shepard, Roger N., and Jacqueline Metzler. "Mental Rotation of Three-Dimensional Objects." *Science* 191 (1971): 952–54.

Woolley, Anita W., J. Richard Hackman, Thomas E. Jerde, Christopher F. Chabris, Sean L. Bennett, and Stephen M. Kosslyn. "Using Brain-Based Measures to Compose Teams: How Individual Capabilities and Team Collaboration Strategies Jointly Shape Performance." *Social Neuroscience* 2 (2007): 96–105.

Chapter 5: Sweeping Claims

Roger W. Sperry and His Work

Bogen, Joseph E. "The Neurosurgeon's Interest in the Corpus Callosum." In *A History of Neurosurgery: In Its Scientific and Professional Contexts*, edited by S. H. Greenblatt, 489–98. Stuttgart: Thieme, 1997.

"Joseph E. Bogen." In *The History of Neuroscience in Autobiography*, vol. 5, edited by L. R. Squire, 47–122. Amsterdam: Elsevier, 2006.

Bogen, Joseph E., E. D. Fisher, and P. J. Vogel. "Cerebral Commissurotomy: A Second Case Report." *Journal of the American Medical Association* 194 (1965): 1328–29.

Gazzaniga, Michael S., Joseph E. Bogen, and Roger W. Sperry. "Some Functional Aspects of Sectioning the Cerebral Commissures in Man." *Proceedings of the National Academy of Sciences* 48 (1962): 1765–69.

———. "Laterality Effects in Somesthesis Following Cerebral Commissurotomy in Man." *Neurospychologia* 1 (1963): 209–15.

———. "Observations on Visual Perception After Disconnexion of the Cerebral Hemispheres in Man." *Brain* 88 (1965): 221–36.

Gordon, H. W., Joseph E. Bogen, and Roger W. Sperry. "Absence of Deconnexion Syndrome in Two Patients with Partial Section of the Neocommisures." *Brain* 94 (1971): 327–36.

Slade, Margot, and Eva Hoffman. "Brainwork Is Rewarding Work." *New York Times*, October 11, 1981.

Sperry, Roger W. (1961). "Cerebral Organization and Behavior: The Split Brain Behaves in Many Respects Like Two Separate Brains, Providing New Research Possibilities." *Science* 133 (1961): 1749–57.

———. "Brain Bisection and Mechanisms of Consciousness." In *Brain and Conscious Experience: Study Week September 28 to October 4, 1964, of the Pontificia Academia Scientiarum in Rome*, edited by J. C. Eccles, 298–313. New York: Springer Verlag, 1965.

———. "Brain Research: Some Head-Splitting Implications." *The Voice* 15 (1965): 11–16.

———. "Lateral Specialization in the Surgically Separated Hemispheres." In *Third Neurosciences Study Program*, edited by F. Schmitt and F. Worden, 5–19. Cambridge, MA: MIT Press, 1974.

———. "Mind-Brain Interaction: Mentalism, Yes; Dualism, No." *Neuroscience* 5, no. 2 (1980): 195–206.

———. "Some Effects of Disconnecting the Cerebral Hemispheres." Nobel Lecture, Nobel Prize in Physiology or Medicine 1981, www.nobelprize.org/nobel_prizes/medicine/laureates/1981/sperry-lecture_en.html#.

———. "Science and Moral Priority: Merging Mind, Brain and Human Values." *Convergence*, vol. 4 (Ser. ed. Ruth Anshen). New York: Columbia University Press, 1982.

———. "Consciousness, Personal Identity and the Divided Brain." *Neuropsychologia* 22 (1984): 661–73.

Sperry, Roger W., and Michael S. Gazzaniga. "Language Following Surgical Disconnection of the Hemispheres." In *Brain Mechanisms Underlying Speech and Language*, edited by D. Millikan, 108–21. New York: Grune & Stratton, 1967.

Origins of the Left, Right Theory

Goleman, Daniel. "Split-Brain Psychology—Fad of the Year." *Psychology Today* 11 (1977): 88–90.

Mitzberg, Henry. "Planning on the Left Side and Managing on the Right." *Harvard Business Review* 54 (1976): 49–58.

Life Magazine began publishing its dramatically illustrated five-part series about the brain on October 1, 1971.

Ornstein, Robert E. *The Psychology of Consciousness.* San Francisco: W. H. Freeman & Company, 1972.

Time, "Hemispherical Thinker," about Robert E. Ornstein, July 8, 1974.

Pines, Maya. "We Are Left-Brained or Right-Brained; Two Astonishingly Different Persons Inhabit Our Heads." *New York Times Magazine*, September 9, 1973.

Brain Myths and Urban Legends

Harnad, Stevan, and Horst D. Steklis. "Comment on J. Paredes and M. Hepburn's 'The Split Brain and the Culture-and-Cognition Paradox." *Current Anthropology* 17 (1976): 320–22.

Hines, Terence. "Left Brain/Right Brain Mythology and Implications for Management and Training." *Academy of Management Review* 12 (1987): 600–06.

Milner, Brenda. "Interhemispheric Differences in the Localization of Psychological Processes in Man." *British Medical Bulletin* 27 (1971): 272–77.

Ornstein, Robert E. *The Right Mind: Making Sense of the Hemispheres.* New York: Harcourt Brace & Company, 1997.

Pietschnig, Jakob, Martin Voracek, and Anton K. Formann. "Mozart Effect-Shmozart Effect: A Meta-Analysis." *Intelligence* 38 (2009): 314–23.

Pink, Daniel H. *A Whole New Mind: Moving from the Information Age to the Conceptual Age.* New York: Riverhead, 2005.

Wieder, Charles G. "The Left-Brain/Right-Brain Model of Mind: Ancient Myth in Modern Garb." *Visual Arts Research* 10 (1984): 66–72.

General Background

Kosslyn, Stephen M., and Olivier Koenig. *Wet Mind: The New Cognitive Neuroscience.* New York: Free Press, 1992 and 1995.

Springer, Sally P., and Georg Deutsch. *Left Brain, Right Brain: Perspectives from Cognitive Neuroscience.* New York: W. H. Freeman and Company, 1981 and 1998.

Chapter 6: Interacting Systems

Gregory, Richard L. "The Brain as an Engineering Problem." In *Current Problems in Animal Behaviour*, edited by W. H. Thorpe and O. L. Zangwill. Cambridge: Cambridge University Press, 1961.

Macmillan, Malcolm. *An Odd Kind of Fame: Stories of Phineas Gage.* Cambridge, MA: MIT Press, 2002.

Chapter 7: Four Cognitive Modes

Biederman, Irving, and Maggie M. Shiffrar. "Sexing Day-Old Chicks: A Case Study and Expert Systems Analysis of a Difficult Perceptual-Learning Task." *Journal of Experimental Psychology: Learning, Memory, and Cognition* 13 (1987): 640–45.

Bouchard, Thomas J., Jr. "Genetic Influence on Human Psychological Traits: A Survey." *Current Directions in Psychological Science* 13 (2004): 148–51.

Ericsson, K. Anders, William G. Chase, and Steve Faloon. "Acquisition of a Memory Skill." *Science* 208 (1980): 1181–82.

Kagan, Jerome, and Nancy Snidman. "Early Childhood Predictors of Adult Anxiety Disorders," *Biological Psychiatry* 46 (1999): 1536–41.

Kagan, Jerome. *Galen's Prophecy: Temperament in Human Nature.* New York: Basic Books, 1994.

Plomin, Robert, John C. DeFries, Valerie S. Knopik, and Jenae M. Neiderhiser. *Behavioral Genetics*, 6th ed. New York: Worth Publishing, 2012.

McManis, Mark H., Jerome Kagan, Nancy C. Snidman, and Sue A. Woodward. "EEG Asymmetry, Power, and Temperament in Children." *Developmental Psychobiology* 41 (2002): 169–77.

Saudino, Kim J. "Behavioral Genetics and Child Temperament." *Journal of Developmental and Behavioral Pediatrics* 26 (2005): 214–23.

Chapter 8: Origins of the Modes: Nature Versus Nurture

Biederman, Irving, and Maggie M. Shiffrar. "Sexing Day-Old Chicks: A Case Study and Expert Systems Analysis of a Difficult Perceptual-Learning Task." *Journal of Experimental Psychology: Learning, Memory, and Cognition* 13 (1987): 640–45.

Bouchard, Thomas J., Jr. "Genetic Influence on Human Psychological Traits: A Survey." *Current Directions in Psychological Science* 13 (2004): 148–51.

Committee on Support for Thinking Spatially. *Learning to Think Spatially: GIS as a Support System in the K–12 Curriculum.* Washington, DC: National Academies Press, 2006.

Ericsson, K. Anders, William G. Chase, and Steve Faloon. "Acquisition of a Memory Skill." *Science* 208 (1980): 1181–82.

Ericsson, K. Anders, V. Patel, and Walter Kintsch. "How Experts' Adaptations to Representative Task Demands Account for the Expertise in Memory Recall: Comment on Vicente and Wang (1998)." *Psychological Review* 107 (2000): 578–92.

Hu, Yi, K. Anders Ericsson, Dan Yang, and Chao Lu. "Superior Self-Paced Memorization of Digits in Spite of a Normal Digit Span: The Structure of a Memorist's Skill." *Journal of Experimental Psychology* 35 (2009): 1426–42.

Kagan, Jerome. *Galen's Prophecy: Temperament in Human Nature.* New York: Basic Books, 1994.

Kagan, Jerome, and Nancy Snidman. "Early Childhood Predictors of Adult Anxiety Disorders." *Biological Psychiatry* 46 (1999): 1536–41.

Kagan, Jerome, Nancy Snidman, Marcel Zentner, and Eric Peterson. "Infant Temperament and Anxious Symptoms in School Age Children." *Developmental Psychopathology* 11 (1999): 209–24.

McManis, Mark H., Jerome Kagan, Nancy C. Snidman, and Sue A. Woodward. "EEG Asymmetry, Power, and Temperament in Children." *Developmental Psychobiology* 41 (2002): 169–77.

Saudino, Kim J. "Behavioral Genetics and Child Temperament." *Journal of Developmental and Behavioral Pediatrics* 26 (2005): 214–23.

Wright, Rebecca, William L. Thompson, Giorgio Ganis, Nora S. Newcombe, and Stephen M. Kosslyn. "Training Generalized Spatial Skills." *Psychonomic Bulletin and Review* 15 (2008): 763–71.

Chapter 9: Mover Mode

Bloomberg, Michael, with Matthew Winkler. *Bloomberg by Bloomberg.* New York: John Wiley & Sons, 1997 and 2001.

Bumiller, Elizabeth. "With Attention Money Can Buy, Bloomberg Kicks Off Campaign." *New York Times*, June 7, 2001.

Freedman, Russell. *The Wright Brothers: How They Invented the Airplane.* New York: Holiday House, 1991.

Purnick, Joyce. *Mike Bloomberg: Money, Power, Politics.* New York: PublicAffairs, 2009.

Chapter 10: Perceiver Mode

Dalai Lama. *Freedom in Exile: The Autobiography of the Dalai Lama.* New York: HarperCollins, 1990.

———. "A Great Tibetan Teacher of Mind Training Once Remarked That One of the Mind's Most Marvelous Qualities Is That It Can Be Transformed," *Hindustan Times*, December 31, 2010.

Habegger, Alfred. *My Wars Are Laid Away in Books: The Life of Emily Dickinson.* New York: Modern Library, 2002.

Higginson, Thomas W., ed. *The Complete Poems of Emily Dickinson.* Los Angeles: Mundus Publishing (Amazon Kindle edition), 2010.

Messages and teachings from the Dalai Lama are found at his website, www.dalailama.com.

Chapter 11: Stimulator Mode

Bosman, Julie. "Emphasizing Frugal Tastes, Palin Addresses Clothing Issue." *New York Times,* October 26, 2008.

Healy, Patrick, and Michael Luo. "$150,000 Wardrobe for Palin May Alter Tailor-Made Image." *New York Times,* October 22, 2008.

Hoffman, Abbie. *The Autobiography of Abbie Hoffman.* New York: Four Walls, Eight Windows, 1980 and 2000.

Palin, Sarah. *Going Rogue: An American Life.* New York: HarperCollins, 2009.

Zeleny, Jeff, and Michael D. Shear. "Palin Joins Debate on Heated Speech with Words That Stir New Controversy." *New York Times,* January 12, 2011.

A summary of several polls on Sarah Palin and the 2008 presidential election is found at www.pollingreport.com/wh08.htm.

Chapter 12: Adaptor Mode

A timeline of baseball's steroid scandal is at http://nbcsports.msnbc.com/id/22247395/.

Heymann, C. David. *Liz: An Intimate Biography of Elizabeth Taylor.* New York: Carol Publishing Group, 1995.

Hutchinson, Bill, and Corky Siemaszko. "A-Rod Rocker-Shocker! Fitness Model, Boston Babe on Alex's Roster." *New York Daily News,* July 9, 2008.

Roberts, Selena. *A-Rod: The Many Lives of Alex Rodriguez.* New York: HarperCollins, 2009.

Schmidt, Michael S. "For Rodriguez and Yankees, Another Bout of Disclosure." *New York Times,* March 1, 2010. The initial report was posted on www.nytimes.com on February 28, 2010.

Verducci, Tom. "Rodriguez Trades Shortstop Legacy for Pinstripes, Championship Prospect." *Sports Illustrated,* February 17, 2004.

Weiss, J. M. "Effects of Coping Behavior in Different Warning Signal Conditions on Stress Pathology in Rats." *Journal of Comparative and Physiological Psychology* 77 (1971): 1–13.

———. "Effects of Coping Behavior With and Without a Feedback Signal on Stress Pathology in Rats." *Journal of Comparative and Physiological Psychology* 77 (1971): 22–30.

Chapter 13: Test Yourself

Blazhenkova, Olesya, and Maria Kozhevnikov. "The New Object-Spatial-Verbal Cognitive Style Model: Theory and Measurement." *Applied Cognitive Psychology* 23 (2009): 638–63

Crowne, Douglas P., and David Marlowe. "A New Scale of Social Desirability Independent of Psychopathology." *Journal of Consulting Psychology* 24 (1960): 349–54.

Goldberg, Lewis R. "An Alternative 'Description of Personality': The Big-Five Factor Structure." *Journal of Personality and Social Psychology* 59 (1990): 1216–29, doi: 10.1037/0022-3514.59.6.1216.

Gosling, Samuel D., Peter J. Rentfrow, and William B. Swann. "A Very Brief Measure of the Big-Five Personality Domains." *Journal of Research in Personality* 37 (2003): 504–28, doi: 10.1016/S0092-6566(03)00046-1.

Kosslyn, Stephen M., and William L. Thompson. "Assessing Habitual Use of Dorsal versus Ventral Brain Processes." *Biologically Inspired Cognitive Architectures* (ePub August 9, 2012), 2, 68–76, doi.org/10.1016/j.bica.2012.07.007.

McCrae, Robert R., and Oliver P. John. "An Introduction to the Five-Factor Model and Its Applications." *Journal of Personality* 60 (1992): 175–215, doi: 10.1111/j.1467-6494.1992.tb00970.

Chapter 14: Working with Others

Gleick, James. "Auto Correct This!" *New York Times*, August 5, 2012, Sunday Review section,

Kosslyn, Stephen M. "On the Evolution of Human Motivation: The Role of Social Prosthetic Systems." In *Evolutionary Cognitive Neuroscience*, edited by S. M. Platek, T. K. Shackelford, and J. P. Keenan, 541–54. Cambridge, MA: MIT Press, 2006.

Sparrow, Betsy, Jenny Liu, and Daniel M. Wegner. "Google Effects on Memory: Cognitive Consequences of Having Information at Our Fingertips." *Science* 333 (2011): 776–78.

Wegner, Daniel M., Toni Giuliano, and Paula Hertel. "Cognitive Interdependence in Close Relationships." In *Compatible and Incompatible Relationships*, edited by W. J. Ickes, 253–76. New York: Springer Verlag, 1985.

Woolley, Anita W., J. Richard Hackman, Thomas E. Jerde, Christopher F. Chabris, Sean L. Bennett, and Stephen M. Kosslyn. "Using Brain-Based Measures to Compose Teams: How Individual Capabilities and Team Collaboration Strategies Jointly Shape Performance." *Social Neuroscience* 2 (2007): 96–105.

Index

About the Authors

STEPHEN M. KOSSLYN is founding dean of the Minerva Schools at the Keck Graduate Institute. He was previously director of the Center for Advanced Study in the Behavioral Sciences at Stanford University, where he was also a professor of psychology. And before that, for more than three decades he was on the faculty of Harvard University, where he also served as chair of the Department of Psychology and Dean of Social Science; he was also associate psychologist in the Department of Neurology at the Massachusetts General Hospital. Kosslyn received his BA from UCLA and his PhD from Stanford University, both in psychology. He is the author or coauthor of thirteen other books and over three hundred scientific papers on topics ranging from the nature of visual mental imagery to the neural foundations of cognition. Kosslyn has received the American Psychological Association's Boyd R. McCandless Young Scientist Award; the National Academy of Sciences Initiatives in Research Award; the Cattell Award; a Guggenheim Fellowship; the J-L. Signoret Prize (France); honorary doctorates from the University of Caen (France), the University of Paris Descartes (Sorbonne), and the University of Bern (Switzerland); and election to Academia Rodinensis pro Remediatione (Switzerland), the Society of Experimental Psychologists, and the American Academy of Arts and

Sciences. Kosslyn loves to work but also has two hobbies: bass guitar (which he can play just well enough to accompany tolerant musician friends) and the French language (which he has struggled with since living in Paris for a year).

G. WAYNE MILLER is a staff writer at *The Providence Journal*; a documentary filmmaker; and the author of seven books of nonfiction, three novels, and three short story collections. He has been honored for his work many times, most recently with the 2013 Roger Williams Independent Voice Award from the Rhode Island International Film Festival, and was a member of *The Providence Journal* team that was a finalist for the 2004 Pulitzer Prize in Public Service. Three documentaries he wrote and coproduced have been broadcast on PBS, including *The Providence Journal's Coming Home*, about veterans of the wars in Iraq and Afghanistan. *Coming Home* was nominated in 2012 for a New England Emmy, and won a regional Edward R. Murrow Award. Miller is Visiting Fellow at Salve Regina University's Pell Center for International Relations and Public Policy, where he is cofounder and codirector of the Story in the Public Square program (publicstory.org), and former chairman of the Board of Trustees and now trustee emeritus of the Jesse M. Smith (public) Memorial Library in Harrisville, Rhode Island. He is active in a number of other civic causes. With Yolanda Gabrielle, he enjoys travel and time by the ocean, particularly the New England coast. Visit him at gwaynemiller.com.